# GLOBAL GROUND WATER RESOURCES AND MANAGEMENT

# GLOBAL GROUND WATER RESOURCES AND MANAGEMENT

Dr. R.N. Sahay

Oxford Book Company
Jaipur ▪ New Delhi

*Published By*

Oxford Book Company

**Regd. Off.** "Bony Residency", Gate No. 2
Opp. Tilak Public School, Vishweswariya Nagar,
Gopalpura Road, Jaipur - 302018, India
Ph: 0141 - 2761280, Telefax: 2761381
email - oxfordbookjpr@gmail.com
website - oxfordbookcompany.com

**Delhi Office:** 102, 1st Floor, Satyam House,
4327/3, Ansari Road, Darya Ganj,
New Delhi - 110002, India
Ph: 011-45652440

*Distributed By*
**KEYA INTERNATIONAL PRESS LLP**
E-mail: keyainternationalpressllp@gmail.com

Published in India, 2025



ISBN : 9789391096274



Typeset by Shivangi Computers, Jaipur

Printed and bound in India by Thomson Press Ltd.,
New Delhi

# Contents

# Preface

Until recently, the eastern United States has been blessed with an abundance of water; unlike the arid West, shortages in the East have historically been "rare and short-lived." During the past few decades, however, water has increasingly become scarce, due to recurring droughts and burgeoning urban and suburban populations. A severe drought struck the East Coast in 2002, forcing residents in rural Maine to stop flushing toilets and washing dishes, significantly damaging crops in the Midwest, and leading New Jersey to ban lawn watering. But that drought was only the latest and harshest of several that has hit the region since the 1980s. Rapid population growth and commercial development in and around eastern cities have also contributed to shortages, in some cases leading to interstate disputes.

Scientists predict that erratic precipitation patterns will persist, producing recurring droughts in years to come; continued growth in demand will only exacerbate the consequences of weather shifts. For policymakers, water shortages present three central challenges. First and most obviously, scarcity reminds us of the limits of our natural resources, and forces us to consider conservation measures. Second, scarcity requires that we assess the efficiency of water allocation and usage. Third—especially in the context of water, a universally necessary resource—equity demands that basic needs be met, and that one group of users not be allowed to exclude another group from an essential resource. In recent decades, water shortages have compelled eastern lawmakers to face these three issues—conservation, efficiency, and equity—directly. The result has been growing dissatisfaction with the legal regime that had historically governed the distribution of water rights in the East: common law riparianism. Under riparianism, "an owner of land abutting a waterbody has the right to have the water continue to flow... subject to the equal rights of each owner to make strictly limited use of the water." Riparian landowners, also called riparians, "share the common benefits that arise from adjacency to defined bodies of water." Although riparian law had governed surface water in the East from the time of the first Anglo-American settlers, it came under increasing criticism during the latter half of the twentieth century. From a conservation standpoint, riparianism failed to protect against excessive diversions by riparians, or to ensure minimum stream flows for the public. From an efficiency standpoint, riparian rights were inherently uncertain because they were correlatively defined, and thus could shift over time as neighbouring users and uses changed; uncertainty, in turn, inhibited investment and prevented the development of markets in transferable water rights. And from an equity standpoint, riparianism was accused of favoring private agrarian interests.

The reader of the book will find comprehensive knowledge with suitable examples to this subject.

*—Dr. R.N. Sahay*

# Chapter 1

## Hydropolitics and Equity Measures for Water-sharing Agreements

Because of water's preeminent role in survival from an individual's biology to a nation's economy political conflicts over international water resources tend to be particularly contentious. The intensity of a water conflict can be exacerbated by a number of factors, including a region's geographic, geopolitic, or hydropolitic landscape. Water conflicts are especially bitter, for example, where the climate is arid, where the co-riparians of regional waterways are otherwise engaged in political confrontation, or where the population's water demand is already approaching or surpassing annual supply.

The watersheds of the arid and volatile Middle East provide settings for water conflicts of extreme intensity, in that each of the three major waterways the Nile, the Jordan, and the Tigris-Euphrates systems have elements of all of these exacerbating factors. In fact, as will be discussed below, scarce water resources have already been at the heart of much of the bitter, occasionally armed, conflict endemic to the region. It is of little wonder that Boutros Boutros-Ghali, currently Secretary-General of the United Nations, in the past suggested that a future war in the Middle East may be fought over water.

One can find room for optimism, however, in the fact that the same characteristics of water resources which fuel conflict can, if managed carefully, induce cooperation in an environment of hostility. According to Frey and Naff, "precisely because it is essential to life and so highly charged, water can perhaps even tends to produce cooperation even in the absence of trust between concerned actors." At the heart of water conflict management, is the question 'equity.' A vague and relative term in any event, criteria for equity are particularly difficult to determine in water conflicts, where international legal guidelines are poorly developed. However, application of an 'equitable' water-sharing agreement along the volatile waterways of the Middle East is a prerequisite to hydropolitical stability which, finally, could help propel political forces away from conflict in favour of cooperation.

This chapter explores the question of equity measures for water-sharing agreements in the context of Middle East hydropolitics and is divided into three parts. The next section provides a brief summary of the hydropolitical history of the major waterways of the Middle East, focusing on international instances of hydro-conflict or, alternately and more rarely, hydro-cooperation. Following is a section which offers various measures of water-sharing equity which have been proposed in different settings and academic disciplines, and which may be useful to the process of water conflict resolution.

## Water Conflict and Cooperation ill the Middle East

Living as they do in a transition zone between Mediterranean, subtropical and arid climates, the people in and around the major watersheds of the Middle East have always been aware of the limits imposed by scarce water resources. Settlements sprang up in fertile valleys or near large, permanent wells, and trade routes were established from oasis to oasis. In ancient times, cycles of weather patterns had occasionally profound effects on the course of history. For example, recent research suggests that climatic changes 10,000 years ago, which caused the average weather patterns around the Dead Sea to become warmer and drier, may have been an important factor in the birth of agriculture in the region.

The fluctuating waters of the ancient Middle East have given rise to legend, extensive water law, and the roots of modern hydrology: The flood experienced by Noah is thought to have centred its devastation around the Babylonian city of Ur, submerging the southern part of the Euphrates for about 150 days, while the code of King Hammurabi contains as many as 300 sections dealing with irrigation. The practice of field surveying was invented to help harness the flooding Nile. In addition, the waters of the region were occasionally intertwined with military strategy as, for instance, when Joshua directed his priests to stem the Jordan's flow with the power of the Ark of the Covenant while he and his army marched across the dry riverbed to attack Jericho.

In the centuries since, the inhabitants of the region and the conquering nations which came and went have lived mostly within the limits of their water resources, using combinations of surface water and well water for survival and livelihood. It is in the beginning of this century, as the competing nationalisms of Jews and Arabs began to re-emerge on the ruins of the Ottoman Empire, that the quest for resources took on a new and vital dimension. In the years that followed World War I, the location of water resources influenced the boundaries first between the British and French mandate powers which acquired control over the region, then between the states which developed subsequently. The Zionist border formulation for a 'national home' presented at the Paris Peace Talks in 1919, for example, was determined by three criteria: historic, strategic, and economic.

Economic considerations were defined almost entirely by water resources. The entire Zionist program of immigration and settlement required water for large-scale irrigation and, in a land with no fossil fuels, for hydro-power. The development plans, and the boundaries which were required, were, "completely dependent" on the acquisition of the "headwaters of the Jordan, the Litani River, the snows of Hermon, the Yarmuk and its tributaries, and the Jabbok."

National interests along the Nile likewise were being informed by water resources development at the same time. In the early 1900's, a relative shortage of cotton on the world market put pressure on Egypt and the Sudan, then under a British-Egyptian condominium, to turn to this summer crop, requiring perennial irrigation over the traditional flood-fed methods. The need for summer water and flood control drove an intensive period of water development along the Nile, with proponents of Egyptian and Sudanese interests occasionally clashing within the British foreign office over whether the emphasis for development ought to be further upstream or down.

The most extensive scheme for comprehensive water development along the Nile, proposed in 1920 and now known as the Century Storage Scheme, included a storage facility on the Uganda-Sudan border, a dam at Sennar to irrigate the Gezira region south of Khartoum, and another dam on the White Nile to hold summer flood water for Egypt. The plan worried some Egyptians, and was criticised by nationalists, because all the major control structures would have been beyond Egyptian territory and authority. A Nile Project Commission mediated the immediate conflict in 1920 although no final agreement was reached at the time. In 1925, however, a new commission made recommendations based on the 1920 estimates which would lead finally to the Nile Waters Agreement of 1929. Four thousand

MCM/yr. was allocated to Sudan but the entire timely flow (from January 20 to July 15) and a total annual amount of 48,000 MCM/ yr. was reserved for Egypt.

Between the World Wars, water became the focus of the greater political argument over how to develop the budding states around the Jordan watershed, particularly Israel and Jordan, and what the 'economic absorptive capacity' would be for immigration. Development plans included the Ionides (1939) plan, a British study which suggested that water would be a limiting factor for any additional immigration to Palestine, and the Lowdermilk (1944) plan which suggested in contrast that, with proper water management, resources would be generated for four million refugees in addition to the 1.8 million Arabs and Jews living in Palestine at the time. British policymakers came down on the side of the Ionides Plan, invoking 'economic absorptive capacity' to limit Jewish immigration and land transfers for the duration of World War II.

As the borders of the new states were defined, sometimes by warfare, in the 1950s and 1960s, each country began to develop its own water resources unilaterally. On the Jordan River, the legacy of the Mandates, and the 1948 war, was a river divided in a manner in which conflict over water resource development was inevitable. By the early 1950s, Arab states were discussing organised exploitation of two northern sources of the Jordan the Hasbani and the Banias. The Israelis also made public their All Israel Plan, which included the draining of Huleh Lake and swamps, diversion of the northern Jordan River and construction of a carrier to the coastal plain and Negev Desert the first out-of-basin transfer for the watershed.

Jordan, in 1951, announced a plan to irrigate the East Ghor of the Jordan Valley by tapping the Yarmuk. At Jordan's announcement, Israel closed the gates of an existing dam south of the Sea of Galilee and began draining the Huleh swamps, which lay within the demilitarised zone with Syria. These actions led to a series of border skirmishes between Israel and Syria which escalated over the summer of 1951. In July, 1953, Israel began construction on the intake of its National Water Carrier at the Daughters of Jacob Bridge north of the Sea of Galilee and in the demilitarised zone. Syria deployed its armed forces along the border and artillery units opened fire on the construction and engineering sites. Syria also protested to the U.N. and, though a 1954 resolution for the resumption of work by Israel carried majority, the USSR vetoed the resolution. The Israelis then moved the intake to its current site at Eshed Kinrot on the northwestern shore of the Sea of Galilee.

Against this tense background, President Dwight Eisenhower sent his special envoy Eric Johnston to the Middle East in October 1953 to try to mediate a comprehensive settlement of the Jordan River system allocations. Johnston's initial proposals were based on a study carried out by Charles Main and the Tennessee Valley Authority (TVA) at the request of the U.N. to develop the area's water resources and to provide for refugee resettlement.

The major features of the Main Plan included small dams on the Hasbani, Dan and Banias, a medium size (175 MCM storage) dam at Maqarin, additional storage in the Sea of Galilee, and gravity-flow canals down both sides of the Jordan Valley. The Main Plan excluded the Litani River and described only in-basin use of the Jordan River water, although it concedes that, "it is recognised that each of these countries may have different ideas about the specific areas within their boundaries to which these waters might be directed." Preliminary allocations gave Israel 394 MCM/yr., Jordan 774 MCM/yr., and Syria 45 MCM/yr.

Both Israel and a united Arab League Technical Committee responded with their own counterproposals, and Johnston worked until the end of 1955 to reconcile these proposals in a Unified Plan amenable to all of the states involved. In the Unified Plan, Johnston accomplished no small degree of compromise. Though they had not met face to face for these negotiations, all states agreed on the need for a regional approach. Israel gave up on integration of the Litani and the Arabs agreed to allow out-of-basin transfer. The Arabs objected, but finally agreed, to storage at both the Maqarin Dam and the Sea of Galilee so long as neither side would have physical control over the share available to the other. Israel objected, but finally agreed, to international supervision of withdrawals and construction. Allocations under the Unified Plan, later known as the Johnston Plan, included 400 MCM/yr. to Israel, 720 MCM/yr. to Jordan, 132 MCM/yr. to Syria and 35 MCM/yr. to Lebanon.

The technical committees from both sides accepted the Unified Plan but forward momentum died out in the political realm, and the Plan was never ratified. Nevertheless, Israel and Jordan have generally adhered to the Johnston allocations and technical representatives from both countries have met from that time until the present two or three times a year at "Picnic Table Talks," named for the site at the confluence of the Yarmuk and Jordan Rivers where the meetings are held, to discuss flow rates and allocations.

To the west at the same time, hydrologic developments along the Nile were caught up in the nationalist movements of both Egypt and Sudan. The Aswan High Dam, with a projected storage capacity of 156,000 MCM/yr., was proposed in 1952 by the new Egyptian government, but debate over whether it was to be built as a unilateral Egyptian project or as a cooperative project with Sudan kept Sudan out of negotiations until 1954. The negotiations which ensued, and carried out with Sudan's struggle for independence as a back-drop, focused not only on what each country's legitimate allocation would be, but whether the dam was even the most efficient method of harnessing the waters of the Nile.

Sudan, hoping to guarantee its own future water needs, advocated the planning envisioned in the Century Storage Scheme, with its emphasis on upstream controls, and balked at Egyptian desires for water additional to its 1929 allocations. Egypt based these new allocations of 62,000 MCM/yr. of the river's projected net discharge of 70,000 MCM/yr. on the basis of the "primary needs" of its larger population and on the lack of any other water supplies. Negotiations were broken off and relations threatened to degrade into military confrontation in 1958 when Egypt sent an unsuccessful expedition into territory in dispute between the two countries.

Sudan attained independence in 1956, but it was with the military regime which gained power in 1958 that Egypt adopted more conciliatory tone in the negotiations which resumed, and with which the Nile Water Treaty was signed on November 8, 1959. The total recognised water rights for each state of 55,500 MCM/yr. for Egypt and 18,500 MCM/yr. for Sudan are still upheld today. The treaty also provided for a Sudanese water "loan" to Egypt of up to 1,500 MCM/yr. through 1977; for a Permanent Joint Technical Committee to resolve disputes and jointly review claims by any other riparian; and for equal sharing of future increases in the yield of the Nile. Egypt and Sudan agreed that the combined needs of other riparians, at the time still under British rule, would not exceed 1,000-2,000 MCM/yr. Ethiopia, which until then had not been a major player in Nile hydropolitics, served notice in 1957 that it would pursue unilateral development of the Nile water resources within its territory, estimated at 75-85% of the annual flow, and suggestions were made recently that Ethiopia may eventually claim up to 40,000 MCM/yr. for its irrigation needs both within and outside of the Nile watershed. No other state riparian to the Nile has ever exercised a legal claim to the waters allocated in the 1959 treaty.

As each state developed its water resources unilaterally, their plans began to overlap. By 1964, for instance, Israel had completed enough of its National Water Carrier that actual diversions from the Jordan River basin to the coastal plain and the Negev were imminent. Although Jordan was also about to begin extracting Yarmuk water for its East Ghor Canal, it was the Israeli diversion which prompted President Nasser to call for the First Arab Summit in January, 1964, including heads of state from the region and North Africa, specifically to discuss a joint strategy on water.

The options presented at the Summit were to complain to the U.N., divert the upper Jordan tributaries into Arab states, as had been discussed by Syria and Jordan since 1953, or to go to war. The decision to divert the rivers prevailed at a Second Summit in September, 1964, and the Arab states agreed to finance a Headwater Diversion project in Lebanon and Syria and to help Jordan build a dam on the Yarmuk. They also made tentative military plans to defend the diversion project.

In 1964, Israel began withdrawing 320 MCM/yr. of Jordan water for its National Water Carrier and Jordan completed a major phase of its East Ghor Canal. In 1965, the Arab states began construction of their Headwater Diversion Plan to prevent the Jordan headwaters from reaching Israel. The plan was to divert the Hasbani into the Litani in Lebanon and the Banias into the Yarmuk where it would be impounded for Jordan and Syria by a dam at Mukheiba. The diversion would divert up to 125 MCM/yr., cut by 35% the installed capacity of the Israeli Carrier, and increase the salinity in the Sea of Galilee by 60 ppm. In March, May, and August of 1965, the Israeli army attacked the diversion works in Syria. These events set off what has been called "a prolonged chain reaction of border violence that linked directly to the events that led to the (June 1967) war. Border incidents continued between Israel and Syria, finally triggering air battles in July 1966 and April 1967, and finally to all-out war in June 1967.

In the territorial gains and improvements in geostrategic positioning which Israel achieved in the war, Israel also improved its "hydrostrategic" position. This was largely due to its acquisition of the Golan Heights. With the Golan Heights, it now held all of the headwaters of the Jordan, with the exception of a section of the Hasbani, and an overlook over much of the Yarmuk, together making the Headwaters Diversion impossible. The West Bank not only provided riparian access to the entire length of the Jordan River, but it overlay three major aquifers, two of which Israel had been tapping into from its side of the Green Line since 1955. Jordan had planned to transport 70-150 MCM/yr. from the Yarmuk River to the West Bank. These plans, too, were abandoned.

In the years since, increased integration of the West Bank and Gaza into the economic and hydrologic networks of Israel has led to increasing hydropolitical tensions. As mentioned above, when Israel took control of the West Bank and Gaza in 1967, the territory captured included the recharge areas for aquifers which follow west and north-west from the West Bank into Israel, and east to the Jordan Valley. The entire renewable recharge of these first two aquifers is already being exploited and the recharge of the third is close to being depleted as well.

In the years of Israeli occupation, a growing West Bank and Gaza population, along with burgeoning Jewish settlements, has increased the burden on the limited groundwater supply, resulting in an exacerbation of already tense political relations. Palestinians have objected strenuously to Israel control of local water resources and to settlement development, which they see as being at their territorial and hydrologic expense. Israeli authorities view hydrologic control in the West Bank as defensive. With about 30% of Israeli water originating on the West Bank, the Israelis perceive the necessity to limit groundwater exploitation in these territories in order to protect the resources themselves, and their wells from saltwater intrusion.

Hydropolitical tensions have not been limited to the Nile and the Jordan basins. In 1975, unilateral water developments came very close to leading to warfare along the Euphrates. The three riparians to the river Turkey, Syria and Iraq had been coexisting with varying degrees of hydropolitical tension through the 1960's. At that time, population pressures drove unilateral developments, particularly in southern Anatolia, with the Keban Dam (1965-73), and in Syria, with the Tabqa Dam (1968-73).

Bilateral and tripartite meetings, occasionally with Soviet involvement, had been carried out between three riparians since the mid 1960's, although no formal agreements had been reached by the time the Keban and Tabqa dams began to fill late in 1973, resulting in decreased flow down-stream. In mid 1974, Syria agreed to an Iraqi request that Syria allow an additional flow of 200 MCM/yr. from Tabqa. The following year, however, the Iraqis claimed that the flow had been dropped form the normal 920 m sup3/sec to an "intolerable" 197 m sup3 /sec, and asked that the Arab League intervene. The Syrians claimed that less than half the river's normal flow had reached its borders that year and, after a barrage of mutually hostile statements, pulled out of an Arab League technical committee formed to mediate the conflict. In May 1975, Syria closed its airspace to Iraqi flights and both Syria and Iraq reportedly transferred troops to their mutual border. Only mediation on the part of Saudi Arabia was able to break the increasing tension, and on June 3, the parties arrived at an agreement which averted the impending violence. Although the terms of the agreement were not made public, Naff and Matson cite Iraqi sources as privately stating that the agreement called for Syria to keep 40% of the flow of the Euphrates within it borders, and to allow the remaining 60% through to Iraq.

By 1991, several events combined to shift the emphasis on the potential for 'hydro-conflict' in the Middle East to the potential for 'hydro-cooperation.' The first event was natural, but limited to the Jordan basin. Three years of below-average rainfall caused a dramatic tightening in the water management practices of each of the riparians, including rationing, cut-backs to agriculture by as much as 30%, and restructuring of water pricing and allocations. Although these steps placed short-term hardships on those affected, they also showed that, for years of normal rainfall, there was still some flexibility in the system. Most water decision-makers agree that these steps, particularly regarding pricing practices and allocations to agriculture, were long overdue.

The next series of events were geopolitical, and region-wide, in nature. The Gulf War in 1990 and the collapse of the Soviet Union caused a realignment of political alliances in the Mideast which finally made possible the first public face-to-face peace talks between Arabs and Israelis, in Madrid on October 30, 1991. During the bilateral negotiations between Israel and each of its neighbours, it was agreed that a second track be established for multilateral negotiations on five subjects deemed 'regional,' including water resources. Although the pace of the peace talks has been at times arduously slow, a venue does finally exist where grievances can be aired and the issue of water-sharing equity can be tackled. This, in itself, may help prevent the some of the pressures that have historically led to some of the most bitter water conflicts in the world.

## Measures of Water-sharing Equity

One problem at the heart of Middle East water conflicts is the fact that there is no internationally accepted definition of water-sharing equity. International water law is ambiguous and often contradictory, and no mechanism exists to enforce principles which are agreed-upon. This section describes some measures of water-sharing equity which do exist, their strengths, and their weaknesses.

### *International Water Law*

According to Cano, international water law did not substantially begin to formulate until after World War I. Since that time, organs of international law have tried to provide a framework for increasingly intensive water use. The concept of a 'drainage basin,' for example, was accepted by the International Law Association in the Helsinki Rules of 1966, which also provides guidelines for "reasonable and equitable" sharing of a common waterway. Article IV of the Helsinki Rules describes the overriding principle:

Each Basin State is entitled, within its territory, to reasonable and equitable share in the beneficial uses of the waters of an international drainage basin. Article V lists no fewer than eleven factors which must be taken into account in defining what is "reasonable and equitable." There is no hierarchy to these components of "reasonable use;" rather they are to be considered as a whole. One important shift in legal thinking in the Helsinki Rules is that they address right to "beneficial use" of water, rather that to water per se.

The International Law Commission, a body of the United Nations, was directed by the General Assembly in 1970 to study "Codification of the Law on Water Courses for Purposes other than Navigation." The general principles being codified include:

1) Common water resources are to be shared equitably between the states entitled to use them, with related corollaries of (a) limited sovereignty, (b) duty to cooperate in development, and (c) protection of common resources.
2) States are responsible for substantial transboundary injury originating in their respective territories.

The problems arise when attempts are made to apply this reasonable but vague language to specific water conflicts. In the Middle East, for example, riparian positions and consequent legal rights shift with changing borders, many of which are still not recognised by the world community. Furthermore, international law only concerns itself with the rights and responsibilities of stares. Some political entities who might claim water rights, therefore, would not be represented, such as the Palestinians along the Jordan or the Kurds along the Euphrates.

International law seeks to develop general principles which can then be applied to specific problems. It is testimony to the difficulty of marrying legal and hydrologic intricacies that the International Law Commission, despite an additional call for codification at the U.N. Water Conference at Mar de Plata in 1977, has not yet completed its task. After twenty years and nine reports, only several articles have been provisionally approved. Even once the details are worked through, the principles would not have the force of law until approved by the U.N. General Assembly. Even then, cases are heard by the International Court of Justice only with the consent of the parties involved, and no practical enforcement mechanism exists to back up the Court's findings, except in the most extreme cases. A state with pressing national interests can therefore disclaim entirely the court's jurisdiction or findings.

## Needs-Based Equity

Many of the common claims for water rights are based either on geography, i.e. from where a river or aquifer originates and how much of that territory falls within a certain state, or on chronology, i.e. who has been using the water the longest. The extreme positions of either definition have been referred to as 'the doctrine of absolute sovereignty' in the first case, stating that a state has absolute rights to water flowing through its territory, and 'prior appropriation' in the second, that is, 'first in time, first in right'.

These conflicting doctrines of geography and chronology clash along all of the rivers of the Middle East, with positions usually defined by relative riparian positions. Down-stream riparians, such as Iraq and Egypt, often receive less rainfall than their upstream neighbours and therefore have depended on river-water for much longer historically. As a consequence, modern 'rights-based' disputes often take the form of upstream riparians such as Ethiopia and

Turkey arguing in favour of the doctrine of absolute sovereignty, with downstream riparians taking the position of prior appropriation.

In many of the Middle East water disputes which have been resolved, however, the paradigms used for negotiations have not been 'rights-based,' either on relative geography or chronology of use, but rather 'needs-based.' In agreements between Egypt and Sudan signed in 1929 and in 1959, for example, allocations were arrived at on the basis of local needs, primarily of agriculture. Egypt argued for a greater share of the Nile because of its larger population and extensive irrigation works. Current allocations of 55,500 MCM/yr. for Egypt and 18,500 MCM/yr. for reflect these needs. Likewise along the Jordan River, the only water agreement ever negotiated (although not ratified), the Johnston Accord, emphasised the needs rather than the inherent rights of each of the riparians. Johnston's approach, based on a report performed under the direction of the Tennessee Valley Authority, was to estimate, without regard to political boundaries, the water needs for all irrigable land within the Jordan Valley basin which could be irrigated by gravity flow. National allocations were then based on these in-basin agricultural needs, with the understanding that each country could then use the water as it wished, including to divert it out-of-basin. This was not only an acceptable formula to the parties at the time, but it allowed for a break-through in negotiations when a land survey of Jordan concluded that its future water needs were lower than previously thought.

Because of its relative success, needs-based allocations have been advocated for the region in recent disputes as well, notably in and around the Jordan River watershed where riparian disputes exist not only along the river itself, but also over several shared groundwater aquifers. Shuval, for example, argues for a minimum baseline allocation between Israel, West Bank Palestinians, and Jordan, based on a per capita allotment of 100 m sup3 /yr. for domestic and industrial use plus 25 m sup3/yr. for agriculture. Adding 65% of urban uses for recycled wastewater, Shuval arrives at a hypothetical 2022 allocation as 950 MCM/yr. for Palestinians, 1,330 MCM/yr. for Jordan, and 1,900 MCM/yr. for Israel. Since the regional annual freshwater supply is only about 2,500 MCM/yr., Shuval also advocates a series of water import schemes and desalination plants to provide the difference between regional supply and future demand.

Wolf likewise advocates a needs-based approach, but considers new sources such as recycled wastewater as separate issues. By planning for total urban needs of 100 m sup3/yr. per person, and extrapolating to the point in the future where all of the basin's 2,500 MCM/yr. has to be allocated first to these needs, in other words when the regional population reaches 25 million, expected in the early part of the next century, Wolf arrives at allocations of 1,000 MCM/yr. for Israel, 1,000 MCM/yr. for Jordan, 300 for the Palestinians on the West Bank, and 200 for those in Gaza.

Although needs-based negotiations have been more successful in practice in the Middle East than rights-based claims, this is not to say agreement has been absolute. As noted above, agreement along the Nile has included only two of the nine riparian states Sudan and Ethiopia, both minor contributors to the river's flow. No other state riparian to the Nile has ever exercised a legal claim to the water allocated in the 1959 treaty. The notable exception, and the one which might argue most adamantly for greater sovereignty, is Ethiopia which contributes between 75-85% of the Nile's flow. Political complexities have likewise hindered ratification of water-sharing agreements along both the Jordan and the Euphrates Rivers, with upstream versus first-user arguments still being prevalent. Success has often depended on how well the negotiating strategy coincided with the political complexities of the region.

## Economic Equity

One lately emerging principle incorporated into water conflict resolution is the allocation of water resources according to its economic value. The idea is that different uses and users of the water along a given water way may place differing values on the resource. Therefore, equitable water-sharing should take into consideration the possibility of increasing the overall efficiency of water utilisation by re-allocating the water according to these values. This principle alone may not be accepted as equitable by the parties involved. However, inclusion of economic aspects in water resource allocation may enhance better cooperation and future collaboration in joint projects in the region of concern.

### *Central Planning vs. Market Approaches*

Allocation according to the economic value of water has usually been demonstrated using two approaches. The long-standing approach assumes a central planning authority who knows what is best for society a "social planner" who views the region as one planning unit. The social planner maximises regional welfare subject to all available water resources in the region and given all possible water utilising sectors. In some instances the social planner (government) also includes preferences (policy).

Second approach is the "water market" approach which employs the market mechanism to achieve an efficient allocation of scarce water resources among competing users.

Examples of these approaches can be found in several studies. Howe and Easter derived necessary conditions for economically efficient interbasin water transfer in the U.S Additional studies considered institutional and economic aspects of international cooperation for interbasin development. Goslin examined the economic, legal and technological aspects of the Colorado River Basin allocation between the US riparian states and Mexico. Krutillas analysed the economics of the Columbia River Agreement between the US and Canada. LeMarquand has developed a framework to analyse economic and political aspects of a water basin development. And Haynes and Whittington suggested a social planner solution for the entire Nile Basin.

Recent studies have questioned the equity and justice associated with market allocations. The conclusion from these studies is that economic considerations alone may not provide an acceptable solution to water allocation problems, especially to solve water allocation disputes between nations. While the social planner and the market approach may provide a unique solution to a problem of regional water allocation, they suffer several drawbacks that may affect the efficiency and the acceptability of the proposed solution. The social planner approach assumes that all social preferences are known and incorporated into the regional objective function. This of course might not be the case, especially when dealing with regional water allocations that involve many countries with cultural differences and preferences.

The market approach assumes the existence of many parties in the region, each acting independently, so that the market price for water reflects its true value for each party. If, in that market, one party's decision does not affect the outcome of other individuals, then the self-interest of the parties lead to an efficient outcome for the whole region. In the case of water, one party's decision may affect another party's outcome, creating what is called an externality or, third party effect. If the externality effect (cost) is not included in the supply curve of water, the market mechanism collapses. This introduces inefficiency into the system and results in what economists call "market failure." In the case of water (in a water basin), the externality effects might be multidirectional. This is particularly true for water basins shared by more than one country, and for water used for more than one purpose. Also, water allocation problems are not exactly similar to market setups that we are familiar with (e.g., the market for cars), because they are characterised by a relatively small number of agents with different objectives and water-related perspectives.

### Game Theory

Game theory is an approach that allows the incorporation of economic and political aspects into a regional water sharing analysis with a relatively small number of participants, each with different objectives and perspectives. The principles of game theory are not discussed here in detail, but can be found elsewhere. For the game theorist, the dichotomy of whether two riparian states or political entities work unilaterally, in pursuit of only their own goals or, alternatively, work cooperatively in pursuit of regional goals, is recognisable as a familiar two-player, two-strategy game. In the language of the theory, unilateral water resources development might be referred to as a "defection" strategy while working together is referred to as a "cooperation" strategy. Each player chooses a strategy depending on such factors as regional geopolitical relations, relative levels of economic development, and riparian position.

For two water basins within the same political entity, with clear water rights and a strong government interest, the game may resemble a what is known as a "Stag Hunt," where mutual cooperation is the rational strategy. Between somewhat hostile players, either within a state but more often internationally, the game becomes a "Prisoner's Dilemma," where, in the absence of strong incentives to cooperate, each player's individual self-interest suggests defection as the rational approach. One example of this might be the Nile basin. In cases of high levels of hostility, game of "Chicken" can develop, with each player competing to divert or degrade the greatest amount of water before the opponent can do the same. The southern-most part of the Jordan River might be used as an example of riparians playing a game of "Chicken," with Syrian, Jordanian, and Israeli unilateral diversions all impeding basin-wide cooperation.

To cooperatively solve the problem of water allocations within the water basin, the parties involved should realise some mutual benefit that can be achieved only through cooperation. In cases of cooperation, each party needs to voluntarily participate, and accept the joint outcome from the cooperative project. Once a cooperative interest exists, the only problem which remains to be solved is the allocation of the associated joint costs or benefits. For a cooperative solution to be accepted by the parties involved, it is required that (a) the joint cost or benefit is partitioned such that each participant is better off compared to a non-cooperative outcome; (b) the partitioned cost or benefit to participants are preferred in the cooperative solution compared to sub-coalitions that include part of the potential participants, and (c) that all the cost or benefit is allocated.

The economic literature dealing with application of game theory solutions does not provide many examples of regional international water sharing problems. Rogers applied a game theory approach to the disputed Ganges-Brahmaputra sub basin that involves different uses of the water by India and Pakistan. The results suggest a range of strategies for cooperation between the two riparian nations which will result in significant benefits to each. In a recent paper, Rogers further discusses cooperative game theory approaches applied to water sharing in the Columbia basin between the US and Canada, the Ganges-Brahmaputra basin between Nepal, India and Bangladesh, and the Nile basin between Ethiopia, Sudan and Egypt. In depth analysis is conducted for the Ganges-Brahmaputra case where a joint solution where each country's welfare is better oft' compared to any non-cooperative solution.

Application of metagame theory, which is a nonnumeric method to analyse political (nonrational) conflicts, has been applied to water resources problems by Hipel et al. The resulting outcome of a conflict is a set of strategies most likely to occur and their payoffs to each participant. Becker and Easter have analysed water management problems in the Great Lakes Region between different US states and between the US and Canada. A central planning solution is compared to a game theory solution with the result being in favour of the game theory solution.

Dinar and Wolf, using a game theory approach, evaluate the idea of trading hydro-technology for inter-basin water transfers among neighbouring nations. They attempt to develop a broader, more realistic approach that addresses both the economic and political problems of the process. A conceptual framework for efficient allocation of water and hydro-technology between two potential cooperators provide the basis for trade of water against water saving technology. A game-theory model is then applied to a potential water trade in the western Middle East, involving Egypt, Israel, the West Bank, and the Gaza Strip. The model allocates potential benefits from trade between the cooperators. Main findings are that economic merit exist for water transfer in the region, but political considerations may harm the process, if not block it. Part of the objection to regional water transfer might be due to unbalanced allocations of the regional gains, and part is due to regional considerations other than directly related to water transfer.

The history of hydropolitics along the rivers of the Middle East exemplifies both the worst and the best of relations over international water. While shared water resources have led to, and occasionally crossed, the brink of armed conflict, they have also been a catalyst to cooperation between otherwise hostile neighbours, albeit rarely and secretively. The major barrier to water's role as an agent of peaceful relations is the lack of a widely accepted measure for equitably dividing shared water resources. Many disciplines offer tentative or partial guidelines, including legal equity, needs-based equity, and economic equity. Each of these measures alone, however, cannot incorporate all of the physical, political, and economic characteristics which are unique to each of the world's international waterways.

In this chapter, we have tried to present the current state of equity measurements in water resources in general, and in Middle East water resources in particular. On-going research is attempting both to integrate each of the most useful theoretical approaches into one encompassing set of guidelines, and in quantifying the results for practical application to any water conflict. With the flow of water ignoring political boundaries, and with appropriate measures of water equity eluding disciplinary boundaries, the history of hydropolitics in the Middle East may give one a glimpse into a gloomy but probable future for many of the more than 200 international river basins. Without agreed-upon criteria for fair ownership and distribution of such a vital resource, many may come to experience the sentiments of Byron:

*"Till taught by pain, men know not waters' worth."*

## The Valuation of Contaminated Land

The valuation of undeveloped land can be one of the most challenging appraisal assignments. One of the reasons can be the uncertainty of its highest and best use. Another is the uncertainty of the timing of its highest and best use, due to business cycles and other factors. The assignment becomes even more challenging when it is found that the land is seriously contaminated. The contamination might alter both the highest and best use of the land and the time required to obtain the regulatory approvals that are required to develop the site.

A client will often ask an appraiser to simply value the land under the assumption that it has no contamination. In other cases, the client may provide the appraiser with an estimate of the cost of remediating a contaminated site. The expected assumption is that the property be valued by simply subtracting the estimated remediation cost from the value of the acreage, assuming that it has no contamination. Both of these scenarios have occurred because, as Simons (1998) notes, there are few appraisers with training adequate to place a value on a contaminated property "as is".

The following is an example of an actual appraisal problem that was used to create a presentation for a study dealing with stigma. The property is referred to as Property A. A developer wanted to build a shopping centre on Property A, a site that cost $48 million. The purchaser paid $24 million in cash and financed the balance with a $24 million loan. Some contamination was found while soil borings were being done for the foundation. A remedial investigation was subsequently made. The resulting 'approved' cost of cleanup was estimated at $8 million.

Let's say that this $8 million cost is financed by a second trust deed. The appraiser for the transaction valued the property at $40 million, 'as-is,' or subject to the contamination. Many would say that a loss in value due to stigma would also occur. If you were the owner of the second mortgage and the owner of the first mortgage foreclosed, would you pay the $24 million required to protect your interest? If not, you have lost $8 million. You would be buying the property and its contamination for $24 million. How many of you would pay off the $24 million dollar mortgage?

A methodology is needed to quantify the downside risk of the $8 million remedial cost seriously understating what will eventually be the total remedial cost in order to answer questions like these. What about the risk of the land losing value whiles the remedial action is in process? It could well be that the land could lose its highest and best use during this time frame due to the development of other nearby retail sites

Lenders are not the only ones that need to have these risks quantified. Any purchaser of contaminated land needs to quantify the financial risk. There is also a need for accuracy for corporate balance sheets. If a major environmental liability is not recognised, the value of an asset can be significantly inflated. As Guntermann notes (1995), there is a significant amount of literature that documents reductions in property value around toxic, chemical and solid waste landfills and an emerging literature on stigma-related damage. The literature referenced in this study was confined to that dealing with the valuation of land that is itself contaminated, not other real estate that is near a contaminated site.

The seminal article by Patchin (1988) provided some examples of how he has valued contaminated property, including one example of how he valued contaminated land - the topic of this study. In the example, he included the cost of cleanup as a line item number and did not discuss the uncertainty of the estimate. The example also assumed that the land lost its highest and best use, since technological constraints precluded remediation to the point where its prior highest and best use would have been allowed. Patchin noted that when some appraisers found that a property was contaminated, they assumed that it was unmarketable and therefore that it was worthless. Patchin disagrees by saying, "the extent and nature of the contamination are the crucial factors in estimating the after value of a contaminated property."

In his second article, Patchin (1991) dealt specifically with the concept of stigma that he broadly defined as a loss in value beyond the cost to cure the contamination itself. He went on to list stigma as being caused by the following factors:

1. Fear of Hidden Cleanup Costs
2. The Trouble Factor
3. Fear of Public Liability
4. Lack of Mortgageability
5. Residential vs. Commercial
6. How Clean is Clean?

Patchin discussed the uncertainty in cleanup cost under the first item above and quotes Houston (1989) as saying that "If the estimated cleanup cost is 'X',the typical buyer will deduct '2X' from his purchase offer. Both Patchin and Wilson describe each contamination problem as being as unique as a fingerprint and Wilson (1992) stated that a consultant should provide a variety of costs. These are the most probable cost, the expected cost and a probable cost range. He gave an example of the cost range from a current assignment. Wilson quoted costs that ranged from a low of $5.5 million to a high of $63 million.

The second issue refers to the return that should be allocated to an entrepreneurial effort that is required to deal with the contamination. An allocation for entrepreneurial profit can be readily made by the use of a developmental approach to value. The third item refers to the possibility that soil contamination will seep onto another owner's property. The likelihood of this seepage happening can be assessed by the use of soil borings from a remedial investigation and can also be quantified.

In the discussion of the fourth factor, Patchin provides an example of a situation in which a property owner cannot get financing for new construction until the remediation has been completed. The prospective future value at the time of completion of the remediation is discounted for the five years of delay required to clean the site.

Healy and Healy (1992) and Kinnard (1996) surveyed lenders to find out how decisions on lending are affected by contamination issues. Both surveys found that one of the greatest concerns on the part of lenders was groundwater contamination. Kinnard devised a survey that shows 35.6% of lenders and 25% of investors avoid property with known soil contamination. The survey results also show that 45.8% of lenders and 41.1% of investors avoid property with known groundwater contamination. Soil contamination can lead to groundwater contamination in some situations.

With regard to the fifth factor, he states that market reactions to contamination differ according to whether the property is residential or commercial. And the sixth factor deals with the uncertainty of the level of cleanup that is required. The cleanup level is expressed as allowable levels of risk to the health of the nearby population, because of the contamination that remains in the soil after remediation.

Mundy (1992a) referenced Patchin's latter article and quotes seven criteria from another reference used to evaluate and determine the degree of stigma. Mundy says that stigma results from perceptions of uncertainty and risk. He noted that it may be relatively easy to quantify a simple contamination problem, but that as the complexity of the contamination increases, the level of uncertainty and perceived risk rises. Mundy then goes on to differentiate between real and perceived risk. Perceived risk is the risk envisioned by those who do not understand the true risk of the contamination.

Perceived risk could account for the differences that Patchin observed in stigma in residential vs. commercial properties. Very few homeowners are able to quantify environmental risk. Most buyers of commercial properties have environmental site assessments prepared. A buyer of a commercial property that did not have a study done to quantify the risk of contamination is not well informed.

Both Mundy (1992b) and Wilson (1994) discuss the variation in value (or lack of) at varying points in time. It is acknowledged that there may be no market for a contaminated property (and no value) until the seriousness and extent of the contamination is quantified. The contention is that a property increases in value while the site is being remediated and that the stigma may eventually disappear. Mundy (1992b) and Chalmers and Roehr (1993) focus on quantifying the loss in value of improved property by using cash flow analysis that provides for a remedial cost estimate over a ten year or so holding period. The discount rate that is used is supposed to reflect stigma by accounting for the "increased risk associated with the contaminated property," which the authors say is difficult to justify.

Lizieri, Scarlett and Finlay (1995) surveyed market participants in the United Kingdom in order to find out how they were dealing with the issue. They found a range of valuation methods that were used. The methods "range from their traditional all risks yield approaches and direct application of costs, through yield decomposition techniques to explicit discounted cash flow scenario modelling approaches and real option pricing models." They concluded that "the property industry is well able to identify problems, but is some way from establishing solutions."

Dixon (1995) concludes from his study of contamination issues in America that a systematic approach to valuation / appraisal of contaminated land is essential. Dixon refers to Wilson (1992) and his contention that a valuation "team" approach consisting of experts in such areas as appraisal and engineering be used and that an appraiser should be the manager of this team. Dixon also provides a quote from

Campanella (1995) who sees the appraiser as a synthesizer of all relevant information, including remediation costs.

The appraisal professional is both the beginning and the end of the valuation process, providing the unimpaired value opinion from which all of the adjustments to value are made and providing the market place insight and technical value adjustment skills required to arrive at the impaired value opinion.

Weber (1996) agreed with Wilson and Campanella and described how the techniques used to quantify environmental risk to human or other biologically valuable receptors could also be used to quantify the financial risk of the contamination. This work described the remedial investigation and feasibility study that is required in the United States for a seriously contaminated site.

It also addressed the uncertainty that any one remedial method would be chosen and the Applicable and Relevant and Appropriate Requirements (ARARs) that govern which method is chosen. More recently, Chalmers (1996) notes that the quantification of risk in light of particular, site-specific factors is the most challenging part of the valuation problem. Chalmers states that risk analysis and quantification must begin with the perceptions of market participants. If the potential buyers of a site are well informed, perceived risk would equal real risk, as defined by his earlier article.

Jackson, Dobroski and Phillips (1997) note that all real estate has risk, but the type of risk that keeps real estate from selling is the risk that cannot be defined or quantified. They have concluded that sellers, lenders and buyers need to understand all of the risks in contaminated real estate and that it is necessary to quantify the risks associated with a specific property. He states that this risk is "always dependent on that property's unique environmental, locational, market, regulatory, and other characteristics." This study deals specifically with a methodology for quantifying the above-described risk.

## A Thesis for Investigation

A number of consultants, e.g., Patchin (1988), Wilson (1992, 1994) and Roddewig (1996) have taken issue with the use of comparable sales for the valuation of contaminated land. This opinion is understandable since each site is unique relative to a number of attributes. References on hydrology and other soil sciences, such as Dragun (1988), Fetter (1994) and Freeze and Cherry (1979) show that, even if two sites otherwise identical (above grade) could be found with contamination, the type and extent of the contamination, the difference in subsurface conditions and proximity to an underground water supply could result in one site having one thousand times the environmental, and hence financial, risk of the other site. Travis and Doty (1990) note that some sites may not be capable of remediation.

Since comparable sales are not likely to be useful in valuing sites with a significant contamination problem, a different valuation thesis will be investigated. The thesis is that a development model that would take into consideration the uncertainty of the remedial cost and prospective future value upon completion of the remedial effort would provide a more reliable estimate of value. The model would make use of Monte Carlo sampling to quantify the financial risk of successfully remediating a site. In order to illustrate this point, let us assume that three sites at three corners of the same intersection have all been contaminated. One from a gas station, a second from a dry cleaner and the third from a car repair business.

Assume that Property A, the site discussed in the Introduction, is located at the northwest corner of the hypothetical intersection of X and Y streets. Property B is located at the northeast corner; Property C, at the southwest corner of the same intersection. These sites are identical except for the type and extent of their contamination. Each has a value of $20,000,000 if not contaminated, as indicated by the recent sale of Property D, located at the southeast corner of the same intersection. The remedial costs of the three sites have been simulated, and the resulting frequency distribution has been overlaid on the locations of each property on the map.

Relatively common forms of continuous distributions have been overlaid upon the simulation of each remedial cost. The assumption implicit in the lender's appraisal was that the probable remedial cost for Property A was normally distributed, similar to that of Property C. The risk analysis shows that the distribution of remedial cost for Property A is much more skewed. It is similar to a Weibull or Gamma distribution. Property C is identical to the sites on the other three corners of X and Y, with the exception of the type and extent of contamination. C's contamination is all near the surface. It will cost, at most, $8.6 million to dig it all up, move it all elsewhere, and fill in the excavation.

A geographic information system (GIS) was used, to show that the contamination has seeped further into the soils at Property A. This GIS program can also calculate the environmental risk to humans from polluting the groundwater below the site-the only local source for drinking water. The minimal remediation cost is $8,000,000. If it does not work, a different action with a cost of $20,000,000 will be required. There is still a risk that a $43 million remedial cost might be required to remediate this site. How should the properties be valued?"

### The Data for the Analysis of Property B

The prospects of incurring large financial losses led to the development of a method of quantifying the financial risk involved in acquiring a contaminated site. It applies to sites that have had a preliminary investigation, which concludes that contamination is likely. This financial risk stems from the possibility that a potential buyer of the site might have to pay for very expensive remediation.

The data for the analysis of hypothetical Property B came from actual cost estimates for another "test" property. The reversionary value in the example is also from the other property. The distribution of uncertain cost is the only item from the analysis that follows that is used to illustrate the risk in purchasing Property B.

Property B is representative of a number of sites that contain underground storage tanks and connecting lines for filling and dispensing material. The material could become a hazardous waste if it were to leak from its container. The financial risk stems from the likelihood that a leakage could have occurred anywhere in the underground system. This leakage could result from improperly pouring toxic chemicals down drains in the past. If so, the chemical could endanger the health of the surrounding population and would require removal.

The possible sources of contamination are described in terms of four units:

- Unit I the underground tank and filler/dispenser lines;
- Unit II an oil drain pit;
- Unit III a waste oil tank;
- Unit IV a clarifier.

An engineer performed a Phase II Environmental Site Assessment in order to confirm the presence of contaminants. It revealed oily residue in the surface soils over most of the site. Tests showed that metals and petroleum hydrocarbons were present at a variety of locations and depths. There were three locations where gasoline components were present and there was some hazardous waste in the clarifier unit.

The process requires that each unit be analysed to determine the risk of the buyer having to incur extensive cleanup cost. The risk analysis involved: breaking the in-ground construction into logical components; identifying major events; estimating the likelihood and cost of each event; and deriving a probability distribution of potential costs.

The risk analysis goes beyond traditional single-point engineering estimates of the most likely, optimistic and pessimistic costs. Environmental engineers are surveyed. These engineers develop a range of possible outcomes from the remedial investigation and the cost of each outcome. The experts also provide the probability of incurring each cost. An event tree lists each of the units. The tree shows the possible remedial requirements and the resultant cost of each one.

## The Conceptual Model

The solution to the problem was to structure it as a land development model. The reversion is an estimate of the prospective future value of the site. Starting with Unit I, there is a chance that remediation of the underground tank might be complete in step A by simple excavation. The reversion might occur at Time T=2, if step A in each section resolves all of the potential problems

Chances are that at least one of the units will require steps B or C, etc. If this happens, it will both add an additional "development cost" to the timeline and push the reversion further into the future. This is because regulators will not provide a letter of No Further Action (NFA) until all of the units are satisfactorily remediated.

The probability of each of the steps having to be taken quantifies the "Chances" referred to above. Multiply each cost by its probability to determine the subtotal for each cost. The next step adds the subtotal cost. The bottom line shows the total remedial costs for all of the units, followed by the reversion.

This model simulates the process thousands of times, collecting the present value that results from each iteration. The model takes all of the value estimates and creates a probability distribution of the value. The appraiser uses the range and probability distributions of the major costs that were provided by the engineers and assigns the likelihoods to the potential outcomes for the construction of the event trees. A detailed description of the simulation and its inferences may be noted by reference to the Addendum. A number of papers were presented on business cycles at the first meeting of the International Real Estate Society. Real estate researchers from Australia, Great Britain, Stockholm, and other parts of the world have noted declines in property values resulting from overbuilding as a result of the business cycle. Jeffries (1996), The Royal Institute of Chartered Surveyors (1995) and Born and Pyhrr (1995) noted the problems in assuming constant trends in values. The method described here can also be used separately from a valuation model for improved property that takes into consideration business cycles. In other words, a property could be valued by the use of business cycle analysis as described by Weber (1997) instead of having to rely on the Ellwood Technique for the income approach, which is recommended in a number of articles.

Again, for longer remediation periods, the risk of a differing highest and best use upon completion of remediation should also be considered for the valuation of undeveloped land. Fanning (1994) provides a number of reasons why this market analysis is so important in estimating future values. If another shopping centre had been developed within the trade area of Parcel A (as was actually planned), it could have resulted in Parcel A losing half its value from this factor alone. A probability distribution was also used to estimate the value of the property upon the completion of remediation. A triangular distribution (for the "test" property) with pessimistic, most likely and optimistic estimates were chosen.

### Probabilistic Present Value Analysis

The results of each of the simulations of the remedial process are used in the spreadsheet for a present value analysis that quantifies the uncertainty of the future value. An explanation of the process follows.

### Simulated Cost and Time for Remediation

This section contains a summary of the results of each spreadsheet simulation. The sum of the individual remedial costs for this iteration is labelled Total Remedial Cost.

### Most Likely Value after Remediation

This reversionary value was derived from this iteration of the sampling of the triangular probability distribution that was used to estimate the value of the reversion after remediation. Estimates are made of the most likely, optimistic and pessimistic reversions for cash flow studies. A pessimistic scenario could reflect the loss in value in the event the site loses its highest and best use by the time that remediation has been completed.

### *Remediation Expense*

Each estimated cost of remediation is spread out evenly over the time (in this case, quarters of a year) that is estimated to be required to complete the remediation of each Operable Unit. Again, the time estimate has been tailored to suit the specific outcome of the remedial sampling of the tasks to be done.

### *Real Estate Tax*

This annual expense is estimated based upon the results of the most likely value estimate after remediation. It also has to be allocated over time based upon the periods that are chosen for the analysis. Other items such as entrepreneurial profit could be structured appropriately and deducted over time in this section of the model.

### *Cash Flow*

The cash flow analysis discounts (at a 3% quarterly discount rate in this example) the cash flow after total remedial cost, real estate taxes, etc., and the reversionary value.

### *Estimated Value "As-Is"*

This value indication is based upon the results of this iteration of the remedial process. The value indications for all 10,000 iterations of the spreadsheet have been plotted in the frequency distribution. The resulting mean value of the "test" property prior to the remediation amounts to about $1,700,000.

### *Calculating Stigma*

The model can also be used to quantify stigma, if stigma is defined to include the financial risk of remediation. An appraiser can communicate the financial risk of a site as part of the process of confirming sales, once the risk of cost overrun is quantified by a probability distribution. Reference to the probability distribution of remedial cost for Property C shows that it will cost, at most, $8,600,000 to remediate this site due to the complete removal of the contamination. Survey research can be used to find out what level of risk potential buyers would take to remediate this site in light of the certainty of its profit potential if redeveloped. If, for example, it was found that the most likely buyers wanted a 95% certainty of covering remedial cost, it is a simple process to find the cost that will provide this level of certainty by finding the point on the X-axis that represents 95% of the area under the leftmost portion of the curve.

Let us say that $8,400,000 is the cost on the X-axis that corresponds to a 95% confidence level. Stigma would be reflected by the difference between $8,400,000 and the most likely remedial cost estimate of $8,000,000. It amounts to $400,000, which is only 5% of unimpaired land value. Application of the risk analysis to Property B shows that mean remedial cost is just $35,000. A 95% confidence level for remedial cost here would only suggest stigma amounting to $20,000, or .1% of the value of a $20,000,000 site. If the site were only worth $20,000 after remediation, the stigma would amount to 100% of the land value. This exemplifies why the calculation of stigma as a percentage of value is not considered to be an appropriate methodology. Probabilistic analysis can also be applied to Property A to determine how much the value of the property should be discounted over and above the $8 million "most likely" remedial cost in order to satisfy the maximal risk requirements of the market for this specific site. It may be that the probability of having to incur the $43,000,000 cost will be great enough to render the property valueless. At least the risk will be quantified so it can be more readily communicated to others. Potential purchasers can now be well informed about the financial risk of the site and perceived risk will equal real risk by the use of this methodology.

The methodology presented here was described to a speaker from an insurance company that provides insurance for contaminated sites. The speaker noted that the company estimators performed similar analysis in order to determine an appropriate premium when they consider providing insurance for seriously contaminated sites. The technique then passes one of the initial requirements of the Daubert factor, which calls for general acceptance of the technique, although it is no longer a precondition to admissibility. The technique could also be used to calculate stigma for an income-producing property, by quantifying the annual "premium" that should be deducted from income due to the financial risk of contamination.

The valuation of contaminated land can be among the most challenging appraisal assignments. One could argue that it requires that the appraiser have a number of atypical skills, such as a good understanding of environmental site assessment and remediation, the use of GIS systems, and significant experience in market and marketability analysis. Weber (1998) provides a number of examples of the use of GIS for dealing with environmental problems and for market analysis as part of the appraisal process. The literature suggests that appraisers are the logical ones to coordinate the efforts of other professionals.

Probability distributions can be a very effective method of communicating risk. An appraisal may be a waste of time and money if a remedial investigation that quantifies the risks involved has not been done. It also does not do an appraiser much good to simply know that a site has been contaminated and that somebody estimated that it would cost $8 million to remediate it. The implications of the cost simulations in the addendum are that percentage adjustments and risk rates are not appropriate for the quantification of stigma, if stigma is defined to include uncertain remedial costs.

Comparable land sales with some form of contamination could be used as part of the valuation process, but they might only show one thing. This is that buyers do not refer to other land sales that also have contamination in order to decide what they will pay for a contaminated site.

A number of environmental engineers such as Swaroop (1987) have provided risk analyses, not unlike the one discussed in this study, to clients who are making purchase decisions. Textbooks such as the one by Covello and Merkhofer (1994) describe the use of Monte Carlo for environmental risk assessment. This study demonstrates that the same techniques can be used for financial risk assessment. The use of these techniques can result in a much better idea of the financial risk involved in the acquisition of contaminated land due to uncertain remedial costs and a better indication of its value.

Finally, Hoyt (1997) notes that the 1993 Supreme Court case known as Daubert has changed the admission requirements of scientific evidence by expert witnesses. He concludes that is imperative that real estate appraisers who testify be sure that their scientific evidence will stand up to the scrutiny of Daubert, or their testimony may be rejected. If there are tools such as the ones described in this study that are acknowledged to be useful in the quantification of environmental risk, it makes sense that appraisers should also use these tools in the valuation of contaminated land.

### A Detailed Description of the Model

Investigations have to be made in order to get a letter from regulators stating that no further action (NFA) is required. Each remedial action for Unit IV and its corresponding probability of being required is as follows.

***Event A:*** There is a 20% chance of getting by with only having to remove the clarifier and thereby receiving approval by the regulatory authorities.

***Event B:*** There is an 80% chance of finding more serious contamination upon removing the clarifier. This would result in having to incur additional cost by having to do more testing. There is an 80% chance of getting an NFA letter after added testing. This is a joint probability, resulting in a .8 times .8 or 64% chance of not having to do more.

***Event C:*** IF event B occurs after A (80% chance), there is a 20% chance of finding dichlorobenzene. If it is found, there is an 99% chance of NFA after an additional small remediation cost. This results in a .8 times .2 times .99, or 15.8% chance of incurring this expense.

***Event D:*** This event assumes that the original testing (80% chance) and the added testing (20% chance) is required and that PCE is found (1%) but it has not migrated to the drinking water (99.9%), resulting in a probability of .8 times .2 times .999, or .1598%.

***Event E:*** Event E is the scenario where everything goes wrong. The probability of PCE being found in the drinking water below the site is .8 times .2 times .1 times .001, or somewhat over one chance in one million.

### Analysis of Units I-IV

This is the section of the spreadsheet that provides each estimate of the total remedial cost that might be required for closure of the site.

### Column Headings

Remedial Uncertainties are the individual events that might be required to be done in order to attain closure. They are followed by the probabilities of the individual remedial tasks being required.

Joint Probabilities lists the probability that the individual task and all of its preceding tasks will be required. The joint probabilities are the ones that are incorporated in the model for sampling. For example, in order for event E under "Unit IV - Clarifier & Leach Field" to happen, it must be preceded by the failure of items A through D to solve the problem. The experts estimate that the chance of this happening is less than 2 in a million. Therefore, it is highly unlikely that the simulation will result in selecting this level of remediation as being required.

Remedial Cost is listed in the next column. It contains cost for each of the remedial tasks that might be required for each of the units.

Cumulative Remedial Cost lists the total remedial cost for each event A-E and all of the events that have to be completed prior to the sampled event. In other words, if the result of the sampling for "Operable Unit I - Tank & Lines" is that steps A through E have to be completed before a letter of NFA is granted, the total cost of the remediation would be $27,360. Expected Cost. This item is calculated by multiplying the Cumulative Remedial Cost by the probability of the cost having to be incurred. It is used to show the most likely cost of remediating the unit from a statistical standpoint.

***Time to Closure:*** It can also be important to estimate how long it could take to remediate the site in the event significant contamination is found. Regulators may be present to test the soil when subsurface components are excavated. Certain types of contamination can take a very long time to remediate and building permits might not be allowed until remediation targets are met. The discounted present value of the land should be considered as part of a possible loss in value resulting from the contamination. The opportunity cost of carrying the land could result in a greater loss in value than the total cost of remediating the contamination. Cumulative Time lists the total time required to remediate the site, in the event more than one of the remedial steps are required. The cumulative time is used to estimate the value of the property at the time remediation has been completed.

### *Row Headings: Results of Sampling (Operable Unit IV as the Example)*

***Simulated Remedial Cost (IV).*** A custom probability distribution can be created to reflect the relative probabilities of each of the remedial uncertainties A-E. The distribution that was created for Unit IV is shown.

It shows the probabilities of outcomes A, B and C. The probabilities of Event D at .16% and Event E, with a likelihood of about 1 in

1,000,000, are so unlikely that they are only noticeable as dots on the X-axis. The statistical sampling process retrieves cost for spreadsheet analysis from cost distributions such as the above that are created for each of the Operable Units.

Simulated Time IV I Cost A-E. Contains the result of sampling the time function for the unit, given the number of steps (Costs) required (two, in this iteration).

Cumulative Cost at the bottom). Refers to the sampling results for the cost of each of the units (I-IV) required to successfully remediate the site.

***Total Cost.*** This item refers to the summation of the sampled cumulative cost of each of the units.

***Time.*** A vector containing the sampled time requirements for remediating each of the units.

***Maximum Time.*** The simulated time required for remediation is the longest time required to remediate any of the units, based upon the outcome of a given iteration. In this iteration, the longest time that is required is 6.05 quarters for Unit III.

### *Simulation of Cumulative Remedial Cost*

A computer-generated spreadsheet is used to simulate the remedial process 10,000 times for each of the units. The mean remedial cost ended up being about $35,000.

### *Implications of the Cost Simulation*

The costs shown above were actual estimates for a contaminated property. They were used to help explain the process and to show that financial risk resulting from remedial uncertainty is relative. The resulting risk is significant if the property without contamination is just worth $50,000, but this risk would only have a nominal effect on the value of a $20 million site. The example shows why percentage adjustments applied to value or risk rates are not considered to be appropriate for quantifying stigma, if contaminated real estate is valued prior to remediation and stigma is defined to include uncertain remedial costs.

# Chapter 2

## Irrigation Water Deliveries versus Consumptive Use

Irrigation of agricultural crops accounts for 80% to 90% of water withdrawals in the semi-arid and arid western United States. Much of the water used for irrigation yields relatively low net returns, so initiatives to meet growing urban and environmental demands by using price incentives are increasingly being discussed and in some cases put into effect. Under the prior appropriation water law doctrine prevalent in the western United States, which relies on the legal ownership of water rights, farmers usually do not pay for irrigation water per se. Rather, they are charged for (some of) the costs of capturing and transporting the water to their farms as well as the operational costs of the irrigation company or district servicing them. However, since the beginning of the 1990s, efforts have been made at both the state and federal levels to encourage agricultural water conservation using a range of measures including water pricing. One example is the Central Valley Project Improvement Act passed by Congress in 1992 that introduced a tiered pricing schedule to be implemented at the irrigation district level.

Another example is the policy adopted in 1996 by the Bureau of Reclamation of the U.S. Department of the Interior which encourages and, in some cases, requires irrigation districts receiving federal reclamation water to incorporate conservation price incentives as part of their conservation plans. A detailed handbook was prepared in 1997 to help agricultural water districts and irrigation organisations develop price incentives to achieve more efficient water use throughout the western United States.

In a river basin context, when analysing competing water demands involving the agricultural sector, it is useful to distinguish among three measures of water use: water withdrawals, deliveries, and consumptive use. Withdrawal measures the amount of water diverted from the surface or ground water source. Delivery refers to the amount of water delivered to the place of use, i.e., the farm. Withdrawal minus conveyance losses equals delivery. Consumptive use is the amount of water that is actually depleted-lost to the atmosphere from evaporation and transpiration from plant and soil surfaces, and embodied in plant products.

**Figure :** District to assess the system of ponds and irrigation water delivery

In field irrigation situations, delivery exceeds consumptive use for several reasons, mainly because of on-farm transit losses and field losses due to the imprecision of the water application practices. For example, in the case of irrigation with open ditch with siphons, in order to assure sufficient water reaches plants at the end of the field, an excess is applied at the beginning. Farmers also may not know the precise amount of irrigation water needed and apply more water than strictly necessary. Furthermore, in some areas, water in excess of consumptive use may be applied to carry salts below the crop root zone. The difference between withdrawals and consumptive use is called return flow. With consumptive use typically amounting to 40% to 60% of deliveries, return flows represent a relatively large portion of deliveries, and in many river basins constitute an important part of the downstream water supply. Water economists have long recognised the importance of considering the different measures of water use in their analyses. Bain, Caves, and Margolis (1966) draw attention to the problem "of placing any emphasis on gross, rather than net, demands for water, since the overall adequacy of water supplies depends on the net consumption occurring in any given use". Nevertheless, most research on irrigation water demand and its price-responsiveness has concentrated on delivery demands and paid little attention to consumptive use demands. This focus is not surprising because delivery is usually the farmers' decision variable, and information on the consumptive use of irrigated crops (and, for that matter, return flows) has not readily been obtained. However, in light of increasing competition between water demands, water policy initiatives such as volumetric charges for irrigation water need to be examined not only with regard to their effect on deliveries, but also on consumptive use and return flows. Although price incentives involving higher costs of irrigation water may change producers' requests for water deliveries, we expect that the amount of water consumed would be relatively less affected. This analysis develops and implements an approach to measuring the relative effects of hypothesised price incentives on deliveries and consumption of irrigation water.

While analyses of the demand for irrigation water use have appeared in the literature since the early 1960s, the elasticity estimates and related policy recommendations differ widely. Some researchers found that farmers are very unresponsive to changes in the price of water. Therefore, they commonly caution against the use of pricing policy to bring about reductions in deliveries, because even for relatively small reductions, large price increases would be necessary-with significant effects on agricultural income and wealth. Other studies indicate a more elastic demand, and their authors conclude that pricing policy would be an effective instrument since it would provide the necessary incentives for farmers to adjust to rising prices by using irrigation water more efficiently. Despite the importance of knowing the responsiveness of farmers to price changes, little systematic study has been carried out on the factors which cause these differing findings.

Several recent papers have used models incorporating water deliveries (or withdrawals) and consumptive use to show that for river basin planning purposes the responsiveness of consumptive use to policy initiatives is often equally or more relevant than is responsiveness of deliveries (or withdrawals). With a focus on the western United States, these studies examine the impact of a number of policy measures such as improved on-farm irrigation efficiency, limits on deliveries, and increases in delivery prices. Using a numerical example, Huffaker and Whittlesey (1995) illustrate that improvements in on-farm irrigation efficiency appear to conserve water by reducing withdrawals, but in reality redistribute water between river and aquifer and do not necessarily change consumptive use.

Focusing on the impacts of limits in agricultural water deliveries, Bernardo et al. (1987) use a farm-level crop simulation/mathematical programming model to show that through better-timed irrigations large decreases in delivery demands may be attained with only marginal reductions in consumptive use and crop yields. Similarly, Burke et al. (2001) explore the effectiveness of limiting deliveries with the goal of conserving water for alternative uses. By linking an on-farm economic decision model with a basin-wide hydrologie model of return flow, they demonstrate that when substitution of technology for water supply is allowed for (in addition to crop switching and land fallowing), the resulting decrease in consumptive use is considerably less than the reduction in deliveries.

Huffaker et al. (1998) analyse the impact of price changes for irrigation deliveries on agricultural water conservation based on theoretical derivations. According to their findings, the common presumption that increasing prices for deliveries leads to agricultural water conservation at levels directly related to the price elasticity of delivery demand is valid only where water unconsumed by crops is irretrievably lost to the river basin.

In that special case, instream flows are reduced by the amount of water applied, and the reduction depends on the price elasticity of delivery demand. However, in the presence of return flows, instream flows decrease by the proportion of water applied which is consumptively used. Farmers would be encouraged by increasing block rates to reduce deliveries but, as improved irrigation technologies are adopted, they would also increase the efficiency with which the reduced deliveries are consumed in crop production. Thus the impact on instream flows is empirically uncertain. Huffaker et al. conclude this uncertainty can only be resolved by assessing water price in terms of consumptive use instead of water delivery.

The model we develop aims to address this issue raised by Huffaker et al. (1998) (and by Bain, Caves, and Margolis more than three decades earlier) by empirically analysing the effect of hypothetical price increases for irrigation water on both the demand for deliveries and the derived demand for consumptive use. Our estimates are based on a two-stage water-crop simulation/linear programming model that was applied to the New Cache La Poudre Irrigation Company (NCLPIC), one of the dozen or so major irrigator-owned cooperatives in the lower South Platte Basin near Greeley, Colorado. Flows from the South Platte River and its tributaries serve the major urban-industrial centres and the most important agricultural region of Colorado. Over 80% of the water withdrawals from the South Platte are used for irrigation.

A linear programming model was chosen to estimate delivery and consumptive use demands since it can be easily adapted to represent numerous options available to farmers for adjusting to increased water prices.

To carefully reflect the yield effects of reduced deliveries, the linear programming model incorporates water-crop production functions computed with a transient-state crop simulation model. An innovative feature of the water-crop production functions used in this analysis is that they show yield and consumptive use not only as a function of the amount of water applied during the season, but also as a function of the number and timing of irrigation events. This allows us to estimate the responsiveness of both deliveries and consumptive use to price changes under a wide range of adjustment options, including changes in irrigated acreage, cropping mix, irrigation technology, and irrigation scheduling.

Based on the model findings, we estimate the price-responsiveness of the demand for deliveries and the derived demand for consumptive use as well as the effect on net farm income, and discuss resulting policy considerations. The quantitative estimates of price elasticities of consumptive use demand are, to the best of our knowledge, the first in the literature to date. A comparison between the price elasticities of delivery and consumptive use demands reveals that consumptive use demand tends to be significantly less responsive to increased water price than delivery demand. This finding implies that price incentives for deliveries are likely to have only limited impacts on basinwide water consumption and would not make much additional water available for competing demands. We also examine the influence of different model assumptions regarding the substitutability of other resources for water on the estimated elasticities, and find that both delivery and consumptive use demand elasticities are quite sensitive to model specification.

## Previous Analyses of Price-Responsiveness of Deliveries

Estimates of the shape of the delivery demand function are commonly based on the use of mathematical programming models, especially linear programming models. Gardner (1983) presents an overview of the studies carried out in California during the 1960s and 1970s. The early studies often intended to show that the delivery demand is more price-responsive than generally believed, and that even for very low prices it is not perfectly inelastic as the U.S. Bureau of Reclamation had claimed in the past. Later studies have constructed subregional or regional demand functions from models of representative farms, and commonly calculated responsiveness by either arc-elasticity estimates along the stepped demand curve or by fitting continuous regression equations to the parametric data. The results typically report either an inelastic estimate for the whole price range considered, or an inelastic estimate for the lower prices and a less inelastic or elastic estimate for the higher prices. The shape of the demand function may be influenced by various factors considered in the model such as the quality of the soil, the length of the furrow, the distinction between surface and groundwater, the product price, and the elasticity of the product demand curve.

During the 1970s and early 1980s, estimates of delivery demands and their shape have been developed with statistical crop-water production functions based on data from field experiments conducted at state experiment stations. These analyses suggest optimal water applications per acre are very unresponsive to changes in price. A reason for this finding is that, while allowing changes in water applications for each of a few selected crops, these studies did not permit shifts in the cropping pattern or provide possibilities for substituting other inputs (e.g., labour or fertilizer) or alternative water application technologies for water.

More recently, econometric studies have used secondary data reflecting actual farmer behaviour. Their estimates tend to be more inelastic than suggested by the mathematical programming models, reflecting, at least in part, the differing assumptions and limitations of the two model types. Econometric models produce positive estimates based on historical data which often show little fluctuation in water prices, while mathematical programming models produce normative estimates based on both historical and synthetic data. The latter can be adapted to represent a wide range of scenarios, and model the impacts of policies for which no historical observations need to exist.

Overall, elasticity estimates vary widely-not only between model types, but also among mathematical programming models. Hartman and Whittlesey (1961), in an early study based on representative farms in Colorado, already noted that in addition to factors such as input and output prices, the kinds of adjustments farmers are allowed to make in the model in response to changes in water supply determine the value of additional water, and thus the shape of the demand curve. Our analysis builds on these findings and explores in more detail the effects of model formulation on the shape of both the delivery demand curve and the consumptive use demand curve.

## Modelling Procedure

The model developed here to measure price-responsiveness of demand for deliveries and consumptive use consists of two parts, an agronomic and an economic model. The data are based on the irrigation practices of NCLPIC in Weld County, Colorado. NCLPIC has relatively senior water rights for river flow, but reservoir water and groundwater from the unconfined shallow alluvial aquifer along the South Platte River are also used for irrigation. The major crops are corn grain, alfalfa, edible dry beans, corn silage, and sugar beets. Farmers almost exclusively use surface technologies for distributing water, including open ditches with siphons, and gated and flexible pipes with and without surge. They typically apply several irrigations per crop and season, each with a more or less fixed amount of water.

With a lack of on-farm data on yield responses to different water supplies for the study area, water-crop production functions were computed using a transient-state crop simulation model. The model was applied to the five main crops grown in the service area of NCLPIC assuming an average-weather year. The main features and results of the crop simulation model are summarised. A detailed description including its input parameters is given by Scheierling, Cardon, and Young (1997).

While crop simulation models employed by economists typically treat the water input as a single value of water applied during the season, our model was adapted to capture the effects of irrigation timing as discrete-input events. The model outputs are watercrop production functions that show the impact of alternative irrigation schedules on consumptive use and yield. Each point represents a relationship between yield (and, implicitly, consumptive use) and the number and timing of irrigation water applications. Using such production functions as an input in the economic model has two advantages. First, it allows us to incorporate irrigation scheduling as a decision variable of farmers faced with increasing water prices; and second, in addition to the conventional relationship between water application and yield, it provides estimates for the policy-relevant variable under consideration i.e., the consumptive use associated with a given number and timing of irrigations.

A deterministic single-period linear programming model was developed to incorporate the physical relationships between water application and yield/consumptive use derived with the crop simulation model. Formulated for the long run, the economic model computes the net return-maximising water applications in the study area in response to hypothesised alternative levels of water price, together with a derived demand for consumptive use. Changes in the output supply from the study area were not expected to be large enough to significantly affect regional product prices. An important assumption is that farmers are well informed not only about prices of inputs and outputs, but also about optimally timed irrigation schedules so that they apply limited irrigation water only in these combinations which result in the highest crop yields. Each crop can be irrigated with any of the five irrigation technologies. As in Caswell and Zilberman (1985), possible yield differences between technologies were not considered. These assumptions generated 245 activities.

Unit net returns were calculated for each activity based on the residual imputation approach (Young, 1996). These net returns are calculated as total revenue per acre (derived by multiplying the yield estimate from the simulation model by the crop price) minus variable costs (labour, materials, fuels, but exclusive of irrigation water costs) and annual overhead (including management) and annualised capital costs (inclusive of a land charge estimated at the value of the land in its next-best use, which is assumed to be the production of nonirrigated winter wheat). The resulting residual was imputed to the water resource and used as a value for net income in the objective function. Volatility and inflation were removed from crop prices by taking a five year average of prices for the period 1989-1993 deflated with the GNP Implicit Price Deflator (Colorado Agricultural Statistics Service). Variable costs were taken from crop budgets for the Nebraska Panhandle and adjusted to Weld County conditions based on the advice of Colorado State University extension agents. Alfalfa establishment costs were amortised over the average stand life in northeastern Colorado. Constraints in the model were defined for land (the service area of NCLPIC, amounting to about 40,000 acres) and water deliveries. The latter include surface water (estimated based on average annual withdrawals from river flow and reservoir rights) and groundwater (estimated based on annual well allotments), and amount up to 120,324 acre-feet available in a typical year. An accounting constraint was formulated to measure consumptive use for each production activity. Other constraints were formulated to reflect the cropping pattern in the study area. Dry beans were limited to 17% of the total irrigated area, considering farmers' risk aversion as a result of highly variable bean prices. Sugar beets were allowed on no more than 7% of the area due to the contractual quotas imposed by the processor. Corn silage may be grown on up to 12% of the area, and alfalfa on up to 27% taking into account the magnitude of demand for fodder from nearby feedlots.

### *Empirical Results*

Computations from the economic model focus on the impact of hypothesised price increases on irrigation water deliveries and consumptive use for the study area. Three scenarios with varying on-farm adjustment possibilities to changes in water price were analysed. Scenario 1 limits adjustments to changes in irrigated acreage and crop mix. It further assumes that the irrigation technology for the whole service area is open ditch with siphons, and the number of irrigations applied to each crop cannot be changed from the one which is optimal at the maximum water availability of 120,324 acre-feet. Growing crops with no irrigation is possible. Scenario 2 is similar to scenario I, except the number of irrigations can be varied and decreased down to zero. Scenario 3 allows for the widest range of adjustments including changes in irrigated acreage, the crop mix, the number of irrigations, and irrigation technologies.

As water price starts to rise from zero, the model indicates that deliveries are most quickly reduced in scenario 3, which allows for the most adjustment possibilities. Farmers in scenarios 1 and 2 with fewer adjustment options are initially much slower in reducing deliveries. As water prices reach very high levels, farmers in scenario I are predicted to be the first to stop irrigating, whereas farmers in scenario 3 continue to demand some irrigation water up to a water price of $292. In scenario 1, where adjustment possibilities are limited to changing irrigated acreage and cropping mix and switching to no irrigation, the model implies consumptive use would decline almost proportionally with deliveries. But consumptive use values decrease only very gradually in scenario 3 because farmers can switch to irrigation technologies with higher application efficiencies without much effect on consumptive use (and yields). At least initially, they can also reduce the number of irrigations without much impact on consumptive use. The number of irrigations for alfalfa, for example, can be reduced to five, or even four, before consumptive use values and yields change significantly. For scenario 3, this stage is predicted to set in when deliveries fall below about 52,000 acre-feet. When prices are so high that deliveries cease, farmers in all scenarios are likely to continue to grow on part of the land crops such as alfalfa which have positive net returns for zero irrigations, and thus consumptive use does not drop to zero.

With an almost linear relationship between deliveries and consumptive use, scenario 1 exhibits a consumptive use demand which, over a wide price range, is similar in shape to the corresponding delivery demand. In contrast, consumptive use demands under scenarios 2 and 3 are predicted to diverge substantially from their respective delivery demands. Specifically, although farmers with more adjustment options respond to rising water prices by decreasing deliveries, their consumptive use values decrease at a much slower rate. At very high prices in all scenarios, consumptive use coming from rainfall and drawdown of soil moisture is estimated to remain at about 20,000 acre-feet.

As water prices start to rise, net incomes in scenarios 1 and 2 fall relatively rapidly, while in scenario 3 changes in net income are initially much less dramatic. As water prices reach very high levels, farmers in all scenarios continue to earn a positive net income since they still have some crop production (and some consumptive use).

Under existing conditions in the lower South Platte Basin where irrigation water costs are quite low, water price increases in the lower ranges were considered to be the ones most relevant for examining the impact on deliveries and consumptive use implied by our analysis. The price changes included comprise an increase from zero to $30, and from $30 to $60 per acre-foot of water. Changes in deliveries and consumptive use, as well as in net income, were examined in both absolute and percentage terms. For a price increase to $30 per acre-foot, scenarios 1 and 2 show decreases in deliveries of slightly less than a third, while the more realistic scenario 3 exhibits a relatively large reduction of nearly 50%. However, for a price increase to $60, deliveries in scenarios 1 and 2 with less adjustment options are much more reduced (by 82% and 78%, respectively) than in scenario 3 where the reduction amounts to 69%.

The predicted changes in consumptive use resulting from the two price increases tend to differ from the changes in deliveries depending on the adjustment options of the particular model formulation. That is, the more adjustment options a scenario allows, the larger the difference between changes in deliveries and consumptive use tends to be. The predicted difference is especially pronounced at a water price increase from $30 to $60 per acre-foot. Under scenario 1, for example, deliveries would be reduced by 82%, while consumptive use would decline by 53%. Under scenario 3, which allows for investments in improved water use efficiency, the reduction in consumptive use would be overwhelmingly smaller than the reduction for deliveries (15% as compared to 69%).

Reductions in net income associated with increased price are estimated to be highest in scenario 1. At $30 per acre-foot, even though deliveries are only reduced by a third, net income decreases by more than half as a result of the pricing policy. Irrigators in scenario 3 still would receive 63% of their initial net income even though deliveries are reduced by half. At $60 per acre-foot, farmers in scenarios 1 and 2 are faced with a net income decrease of 79% and 70%, respectively, while farmers in scenario 3 still make 41% of their initial net income. This is because the potential to substitute capital for expensive water allows profitable production to persist in scenario 3 in contrast to the scenarios with more limited adjustment options.

To further illustrate the different effects of water price increases on deliveries and consumptive use, arc elasticities were calculated for delivery and consumptive use demands. The arc formula computes an elasticity at an average between two points, and allows for easy comparison between different scenarios. In the lowest price range (up to $30 per acre-foot), the demand for deliveries is more inelastic for scenarios 1 and 2 than for scenario 3 with more adjustment options, because initially these scenarios require relatively higher price increases for an adjustment, and thus a change in demand for deliveries, to take place. In the price range between $30 and $60 per acre-foot, the delivery demands for scenarios 1 and 2 become elastic, while the delivery demand for scenario 3 remains inelastic. As prices rise to higher levels, the model suggests that the delivery demands for all scenarios become elastic at some point. But especially for scenario 1, delivery demand tends to become more elastic faster because, as irrigation is ceased on more and more crops and irrigated acreage is given up, deliveries decrease more rapidly toward zero. (Switches to inelastic or perfectly inelastic estimates at some of the higher price ranges are caused by high "steps" in the demand curves, where deliveries are estimated to remain unchanged even though prices continue to rise.)

Consumptive use demand also tends to become less inelastic as water prices start to rise. But again, the predicted effects of increasing water prices are seen to depend on the scenario. The consumptive use demands for scenarios 2 and 3 which allow for more substitution possibilities as water cost increases-tend to be more inelastic than those of scenario 1. Further, consumptive use demands in all scenarios are more inelastic than their respective delivery demands over all price ranges. Thus the findings confirm that the price elasticity of consumptive use demand generally cannot be assumed to equal the price elasticity of delivery demand and, in particular, that the demand for consumptive use is likely to be much less price-responsive than the demand for deliveries.

The results reported provide evidence that elasticity estimates for both delivery demand and consumptive use demand are very dependent on the model framework within which they are derived. In the model formulations where farmers' adjustment options are relatively limited, there is a strong correlation between deliveries and consumptive use. Yet for the more realistically formulated scenarios, the correlation between deliveries and consumptive use is likely to be much weaker. This is because delivery is an input factor which, at least to some extent, can be substituted with other input factors such as management (adaptation in the number of irrigations) and/or capital (change to irrigation technologies with higher application efficiencies). Model formulations incorporating these substitutions indicate that they enable farmers to significantly reduce deliveries in response to price increases; at the same time, they can prevent large reductions in consumptive use, and agricultural production, over a relatively wide price range.

The elasticity estimates are also influenced by the method used to calculate them. In our case, most elasticity values would be different if the endpoints between which the arc elasticities are calculated were changed. Therefore, the emphasis here is not on the particular magnitudes of the elasticity estimates, but on the direction of their change depending on the model formulation, the price range considered, and the focus on either delivery or consumptive use demand. Our findings on the price elasticities of delivery demand are in line with previous results of linear programming models which indicated an inelastic demand for lower prices and a less inelastic demand for higher prices. However, they do not support the common presumption that in the case of an inelastic delivery demand, the use of pricing policy would not be very effective in bringing about reductions in deliveries-because, as the argument goes, even for relatively small reductions, large price increases would be required which in turn would cause large negative effects on income and wealth. Instead, this research suggests, especially for the more realistic scenarios with a range of adjustment options, an inelastic delivery demand does not necessarily imply that deliveries cannot be substantially reduced as the price rises. If in scenario 3 the price increases to $30 per acre-foot, the estimate of the delivery demand elasticity is -0.33 but deliveries would be reduced from 120,324 to 60,584 acre-feet.

The modelling approach presented focuses on farmers' responses to hypothesised increases in irrigation water prices. To empirically examine basin-wide hydrologie and ecological effects of rising irrigation water prices and reduced agricultural deliveries (and withdrawals), a river basin optimisation model would have to be used such as, for example, Booker and Young (1994). Basin-wide effects are likely to include increased instream flows that could help restore river ecosystems. They could also involve improved water quality by leaving more native water in-stream for pollution dilution, and by reducing irrigation return flows with polluting salt, nutrient and pesticide loading.

## Policy Considerations

Our empirical results have important implications for irrigation water pricing policies in river basins where water is withdrawn from surface diversions and/or shallow alluvial aquifers, and where return flows constitute a significant part of the downstream water supplies. This would be the case in Colorado's South Platte Basin where annual surface withdrawals are estimated to amount to about 2.5 times the annual native surface water flows, implying return flows are significant sources of downstream water supplies.

Based on results from the more realistic scenarios, even if delivery demands are price-inelastic, volumetric water charges could bring about large reductions in deliveries while not being very effective in reducing water consumption. They would also have the potential of altering basin hydrology, by reducing the magnitude and changing the timing of return flows. In the South Platte Basin, discharges from the alluvial aquifer (largely fed by irrigation return flows) increase late summer streamflows, which can help meet downstream irrigation demands for extended periods after peak runoff from snowmelt. The aquifer functions in the same manner as a reservoir, with irrigation withdrawals in excess of consumptive use contributing to storage and return flows representing releases. A water pricing policy that would significantly decrease deliveries, and return flows, could negatively impact water users who depend on these return flows downstream. Changes in the return flow regime could also impact groundwater levels and pumping costs. (Reduced return flows would not be a concern in the lower reaches of watersheds where there is no downstream dependence on return flows.)

If the goal is to make additional water available for alternative uses, in many river basins pricing of irrigation water deliveries would be a relatively ineffective policy instrument because, under realistic assumptions regarding potential farmer adjustments, net downstream river flows would not change much over a wide price range. In such hydrologie settings, the focus needs to be on measures that actually reduce consumptive use. Structural measures, such as land leveling or replacing open ditches with underground pipes, which mainly decrease transpiration from soil surfaces, would be one such approach. However, the amount of water involved is likely to be relatively small so that, without subsidies, the economic incentive for any single farmer to invest in such measures is minimal.

Larger quantities of water can be made available by changes in the crop mix, including switching to crops with lower seasonal consumptive use, or reducing irrigated acreage and switching to dryland crops. In the lower South Platte Basin, for example, farmers could switch from feed crops (alfalfa and corn) to vegetables such as carrots, which have a much lower water consumption and potentially generate much higher net income. However, these changes would involve making wholesale modifications in farm operations and entering a more dynamic marketing environment. Thus, they are not likely to occur on a widespread basis. To encourage such crop switches as well as potential transitions to rainfed agriculture or rangeland-which otherwise would only occur at very high water prices-subsidies or outright purchases of water rights would be necessary. Also, current federal subsidies for crops with high water consumption (such as sugar beets) might be reoriented to low-consumption/ dryland crops in order to support increased water use in nonagricultural sectors.

If volumetric charges were imposed on consumptive use rather than on delivery amounts, irrigation water pricing could theoretically be an effective policy instrument for reducing the amount of water consumed. Similarly, the problems surrounding irrigation return flows could be avoided if the surface and ground water rights under the prior appropriation doctrine were converted from withdrawal or delivery to consumptive use amounts. But measuring and monitoring actual consumptive use would still be difficult and costly. Short of these changes, policies regarding water transfers need to emphasize the importance of consumptive use rather than withdrawals and deliveries.

As water scarcity increases in the semi-arid West, it is often suggested that an appropriate pricing policy for irrigation water could reduce agricultural water use by providing stimuli for farmers to use water more efficiently, and make water available for higher valued nonagricultural uses. A number of earlier studies have noted that basin-wide water conservation depends less on changes in agricultural withdrawals and deliveries than on reductions in consumptive use. Our analysis extends this research by modelling both the demand for deliveries and the derived demand for consumptive use for a full range of hypothetical price changes.

Data from an irrigation company in Colorado's South Platte River basin were used as the basis for a water-crop simulation/linear programming approach designed to predict farmers' crop and water management responses to increasing water prices. In the context of the irrigation company, assuming an average-weather year, optimally timed irrigations, and a range of adjustment options, a hypothetical price increase from zero to $30 per acre-foot is estimated to reduce deliveries by half but consumptive use only by about 1%. This finding implies that a water pricing policy, or other methods of encouraging reduced deliveries, while maintaining the agriculture base, is not likely to make nearly as much additional water available for alternative urban or environmental uses as might be expected from a less realistic view of the hydrologic-economie interrelationships in irrigated agriculture. Moreover, a pricing policy would have a relatively large negative effect on agricultural net income and, correspondingly, on capitalised net income in farm land values.

Another finding of the analysis is that the estimated impact of the pricing policy depends on the particular model formulation. If adjustment options are relatively limited, the price-responsiveness of delivery demand and consumptive use demand differ little. But in a more realistic model formulation which also allows for adapting in the long run the number of irrigations and switching to more capital-intensive technologies with higher application efficiencies, the predicted consumptive use demand is significantly less responsive to increasing water prices than the delivery demand.

These results have important implications for potential water pricing policies in river basins where return flows constitute a significant part of the downstream water supplies. Our study finds that in such contexts, attempts to make additional water available for nonagricultural uses by raising irrigation water prices would be relatively ineffective in the long run because it would leave net downstream river flows and/or aquifer storage virtually unchanged. Making significant amounts of additional water available must rest on measures which actually reduce consumptive use. This will be most effectively accomplished by changes to low-consumptive use crops or to nonirrigated agriculture.

## Effects of Threshold Drainage Area

HAZUS (abbreviation for HAZards United States) is a GIS-based natural hazard loss estimation software package developed and freely distributed by the United States Federal Emergency Management Agency (FEMA). The current version of HAZUS, HAZUS-MH (HAZUS-MultiHazard), can model three types of hazards: flooding, hurricanes, and earthquakes. Based on inventory data of buildings, vehicles, transportation systems, utility systems, critical facilities, hazardous materials, agricultural commodities, and demographics, as well as specified hazard parameters, HAZUS locates hazard areas and estimates the potential physical, social, and economic costs based on engineering knowledge of the damage and hazard loss estimates (FEMA 2005). Hazard loss estimation provides a basis for hazard preparation, mitigation strategy development, and response and recovery planning, and it is essential to decision making at all levels of government. Equipped with information about the potential loss, hazard managers are able to better priorities mitigation measures to reduce further disaster losses, and local communities can allocate limited resources more efficiently to responding to, and recovering from, disaster events. This study intends to study how the HAZUS parameter of threshold drainage area affects flood modelling results. A threshold drainage area (thereafter drainage threshold) is a parameter that a user specifies to delineate a synthetic stream network from a Digital Elevation Model (DEM). It is the minimum accumulation area draining into a stream network. Decreasing drainage threshold will generate a more detailed stream network with additional tributaries. Theoretically, the minimum drainage threshold is only limited by the resolution of available DEM data (FEMA 2009a). Past studies of hydrological modelling often regard a drainage threshold as optimal if the accordingly developed drainage network matches the real-world stream network. The realistic drainage threshold may change with local relief, soil, geology, and climate conditions. Nevertheless, stream networks developed in HAZUS are for flood modelling purposes and not necessarily the same as real-world stream networks. Apparently, appropriate drainage thresholds for HAZUS flood modelling largely depend on user's need.

A small drainage threshold delineates small stream reaches and thus a maximum coverage of potential floodplains, though the offshoot floodplains may be spurious. As most hydrological studies aim to use real drainage thresholds to delineate stream networks, literature on the effects of specified threshold parameter on hydrological modelling is limited. Moussa and Bocquillon (1996) and Stepinski and Collier (2004) reported that the total length of stream networks decreases with increasing drainage threshold. Li et al. (2008) found that, as drainage threshold increases, the outflows of basins became slower, the discharge of flood peak decreases, and the basin mean concentration time becomes longer. Liu et al. (2003) indicated that using a large threshold value will result in a longer flow time and a prolonged flow response at the end of the flow path.

Another objective of this study is to investigate the effects of study region size on flood modelling. A HAZUS study region is the area defined by a user in the beginning of analysis to allow HAZUS to prepare inventory and DEM data (for delineating stream networks). Note that users are allowed to select stream reaches of interest as a hazard scenario to conduct floodplain delineation and loss estimation. We expect that different sizes of DEM data will produce different stream networks and, consequently, different flood modelling results. Limited studies have mentioned the effects of study region size (channel network extent) on hydrological modelling. Montgomery and Foufoula-

Georgiou (1993) indicated that channel network extent determines hillslope travel distances and network link lengths; the hydrologic response functions of a catchment will therefore be affected. FEMA (2009b) stated that study region size (thereafter region size) may affect drainage areas for stream reaches at the boarders of the study region and cause discontinuous stream reaches. While a stream network delineated for a larger study region appears to be more accurate than a stream network delineated for a smaller corresponding region, FEMA (2009b) recommended a four-county limitation of region size due to the extensive processing of the associated large amount of data.

As not every HAZUS flood study clarified the used parameters of study region and drainage threshold (e.g., Burton and Cutter 2008), most studies determined the parameters based on reasonable or convenient choices. The study region was set to cover the streams of interest, and the default drainage threshold of 10 square miles in HAZUS was used. Some studies have an area of interest larger than one county and set the study region to one county for computational efficiency. Considering floodplains delineated from more detailed stream networks to be more accurate, Gall et al. (2007) used the minimum drainage threshold of 1 square mile allowed by the initial release of HAZUS-MH. "Considering the tradeoff between stream network detail and computer processing time", Ding et al. (2008) determined an optimal drainage threshold of 5 square miles for a watershed of 287 square kilometres. In delineating floodplains for all counties in Maryland, Joyce and Scott (2005) used different drainage thresholds for individual counties "without creating an overwhelming number of reaches", which are between 4 to 6 sq mi for most counties. Parameters of drainage threshold and region size are artificial factors introduced by HAZUS users that may have significant impacts on flood modelling results. While the appropriate parameters largely depend on characteristics of the study area and analysis needs, understanding potential effects of the parameters helps users to choose optimal parameters for the study area. Identifying trends and variations of the loss estimates due to changes in the parameters is also essential to HAZUS application. This study intends to establish the methodology for performing such an investigation through a case study.

## Flood Hazard Modelling

HAZUS flood modelling involves analyses of flood hazard and flood loss. Flood hazard analysis consists of three stages: stream delineation, hydrologic analysis to determine stream discharge, and hydraulic analysis to determine flood depth and extent. Stream delineation generates a flow direction and a flow accumulation grid from a DEM. Stream networks are then derived comprising grid cells into which more areas than the threshold area drain.

**Figure :** Flood hazard modelling

Hydrologic analysis of stream discharge estimates the discharges at each stream's upstream and downstream nodes for specified flood return periods. The analysis uses default regression equations in HAZUS; the input variables are calculated stream and catchment statistics; and default temperature, precipitation, soil, and land cover parameters are taken into account. Estimated discharges are adjusted by default discharge data at stream gages and by default discharge frequency curves at designated main streams.

Hydraulic analysis of flood depth and extent computes flood elevations at cross-sections of stream channels based on channel morphology and the estimated discharges (for a specified flood return period). These elevations are used to form a flood elevation grid from which a flood depth grid is computed. The flood depth grid and inventory spatial data are input in a flood loss analysis to estimate possible damage to buildings, infrastructure, and key facilities.

Because building inventory data in HAZUS are summed by census block, flooded building counts are estimated by areal-weighting the general building stock within a census block. Note that this areal-weighting approach may overestimate flooded buildings, since buildings are likely to concentrate outside potential flooding areas rather than distributed equally as the model assumes. HAZUS estimates the degree of building damage by default depth-damage curves. Flooding debris and the direct and indirect economic losses are estimated thus:

- For buildings with less than 50 percent damage, the estimates are based on occupancy type and flood depth; and
- For buildings exceeding the 50 percent damage threshold, the building is assumed to be a total loss and will be demolished.

Further, HAZUS estimates shelter needs based on estimated building populations affected by flooding. To assess the effects of HAZUS input parameters of study region size and threshold drainage area, this study focuses on three HAZUS flood modelling outputs total floodplain area, total amount of flood debris, and total economic loss. Flood-generated debris may include building finishes (e.g., dry wall, carpeting, and insulation) and building contents (e.g., furniture and appliances). In severe cases, flood may damage building structure and foundation and produce debris of structural materials (e.g., framing, walls, and exterior cladding) and foundation materials (concrete slab, concrete block, or other foundation). Flood-fighting effort and the floodwater add additional debris, such as sandbags, mud, and sediment. The HAZUS flood debris model only estimates weights of debris generated from building components, including finishes, structure, and foundation; building contents and debris loads from vegetation and sediment are not included.

The direct loss estimates in HAZUS include the costs of building repair and replacement. The indirect economic loss includes the consequential loss from the immediate impact of building damage on a community, such as financial resources needed to recover from damages, financial consequences to the community's businesses due to businesses interruption, and financial costs due to potential job and housing losses (FEMA 2009b). In this study we use HAZUS to model floodplains based on a 100 yr flood return period and compare them with the 100 yr floodplains delineated on FEMA's Flood Insurance Rate Maps (FIRMs). Hundred year floodplains are areas "subject to inundation by a flood that has a 1percent or greater chance of being equalled or exceeded during any given year" (FEMA 2008a). The 100 yr floodplain is a regulatory standard used by federal agencies to administer floodplain management programs; it is also the basis for insurance requirements in the National Flood Insurance Program (NFIP). According to FEMA (2008a), the delineation of floodplains on FIRMs "is based on historic, meteorologic, hydrologic, and hydraulic data, as well as open-space conditions, flood control works, and development." In addition to FIRMs, FEMA also uses Digital Q3 Flood Data. This layer was created by scanning paper copies of effective FIRMs and digitising the floodplain boundaries (FEMA 2008b). It is worth noting that FIRMs may be outdated, particularly if local topography has changed.

HAZUS flood modelling allows for three levels of analysis. Level 1 analysis, which is based on HAZUS default data and settings, allows users to easily implement flood modelling without having extensive engineering knowledge or detailed local data, such as LIDAR derived DEM, field-measured flood-depths, and updated building inventory. However, flood distribution and loss from Level 1 analysis are relatively crude estimates and can have large margins of error (FEMA 2009b), meaning that the estimates may not be suitable for hazard mitigation decision making by local governments. According to FEMA, the estimates "should only be used as initial estimates to determine where more detailed data collection and analyses are warranted, or for studies covering large geographic areas." "This type of output would be appropriate for regional, state or national level analyses" to "provide a benchmark against which progress toward reducing losses due to natural hazards through public policy can be assessed". Level 2 analysis incorporates user-collected local building inventory, elevation, or flood depth data to improve hazard analysis and loss estimate. Level 3 analysis incorporates data and information of a specific hazard event acquired from site-specific engineering and economic studies; the HAZUS default stream discharge functions and depth-damage curves may be modified by technical experts.

This study conducts a Level 1 analysis, which allows us to efficiently implement HAZUS flood modelling for the purpose of assessing the effects of input parameters of study region size and threshold drainage area. HAZUS default elevation data, the 30 m DEM data of the National Elevation Dataset (NED), are used for developing synthetic stream networks. The National Elevation Dataset covers the entire U.S. and is accessible from the seamless U.S. Geological Survey (USGS) web site through HAZUS interface.

Studies evaluating HAZUS flooding modelling accuracy have reached miscellaneous conclusions. Ding et al. (2008) reported underestimated total floodplain areas of approximately 32 percent in a Level 1 analysis. Gall et al. (2007) argued that HAZUS is prone to under-predicting floodplains in upland or undulating terrain because floodplains not connected to a stream, e.g., a lake or pond feature, are not modelled. Contrary to conclusion of Ding et al. (2008) and Gall et al. (2007), URS Group, Inc. (2007) found overestimated floodplain areas in a Level 1 analysis; they claimed the likely reasons for this being the inability to interpret narrow channels from coarse DEMs or ponded flood water due to debris in stream channels. URS Group, Inc. (2007) also reported a 30 percent overestimation of residential loss ostensibly resulting from the even-distribution assumption and area-weighting loss estimation in HAZUS. Supporting the same argument, Ding et al. (2008) found over-estimated flood loss in a Level 1 analysis. Contrary to findings of Ding et al. (2008) and URS (2007), Meyer (2004) reported a six percent underestimation of the total economic loss from building damage in a Level 1 analysis. The different conclusions of HAZUS accuracy studies are likely due to the varied local topography and facility/housing distribution in corresponding study areas.

Our study area is the city of Ozark in Christian county, Missouri, USA, approximately 5 km wide and 10 km long, covering an area of 19.6 square kilometres. Ozark is approximately one kilometre south of the city of Springfield, the major urban centre in southwest Missouri, and approximately 33 km north of the city of Branson, a popular destination for tourists from Missouri and the surrounding states. Highway 65 is the main access mute to Springfield and Branson. As an outlying community within the Springfield Metropolitan Statistical Area, Ozark has experienced rapid growth in recent years. The population of Ozark grew from 4243 in 1990 to 9665 in 2000. In 2007, the estimated population was 17,177. This growth highlights the importance of a flood hazard mitigation plan for the city, which necessitates risk assessment and vulnerability analysis for floodplains. For Ozark, significant flood-prone areas exist in the riparian areas of the Finley River and its tributaries.

Ozark is situated in the Ozark Plateau physiographic region, which is known for a well developed karst topography. A karst landscape has land forms of shafts, tunnels, caves, and sinkholes created by groundwater dissolving sedimentary rock such as limestone. A karst landscape is scenic yet fragile and vulnerable to erosion and pollution. Frequent flooding on Ozark's karst landscape creates compound damage to the natural environment and man-made structures.

To study the effects of drainage threshold on HAZUS flood analysis for the Ozark city, we created a HAZUS study region based on all census blocks within Ozark. The USGS DEM data for the study region were used to develop synthetic stream networks. HAZUS automatically determines the needed DEM boundaries for the study region and extracts the DEM data by first selecting the watersheds (2) that intersect a 2000 m buffer of the study region and then clipping the DEM data to the selected watersheds with a 1,000 m buffer. This approach ensures sufficient DEM data for subsequent flood analysis in HAZUS. We then specified different drainage thresholds for HAZUS to develop synthetic stream networks. A total of 24 drainage thresholds from 0.25 to 6.00 square miles were tested for every 0.25 square mile increase. We then selected stream reaches that geometrically intersect the Ozark city to create the flood scenario.

HAZUS hydrologic and hydraulic analyses were conducted to generate 100 yr floodplain boundaries and associated flood depth grids. HAZUS loss estimates were conducted based on derived flood depth grids and HAZUS default inventory data. We compared the total floodplain area, total flood debris, and total economic loss for different drainage thresholds. The modelled flood plains were also compared with floodplains delineated in FEMA's Digital Q3 Flood Data. Because the selected stream reaches for flood modelling extend outside the boundaries of the city of Ozark, the modelled floodplains from HAZUS cover areas outside the city.

To compare the floodplain areas of Ozark at different drainage thresholds, we clipped the modelled floodplains to the Ozark city boundaries and calculated the clipped floodplain areas. We compared HAZUS-modelled flood debris and economic loss without clipping or other processing, understanding that HAZUS generates debris and loss estimates based solely on inventory data within the defined study region and does not consider inventory data outside the study region (FEMA 2009b).

To study the effects of region size on HAZUS flood analysis for the Ozark city, we created different sizes of study regions enclosing the geographic areas of the city. A total of 20 study regions from a size slightly larger than the Ozark city to a size of two counties (including the local and a neighbouring county) were determined in such a way that the DEM data covered an incremental number of watersheds from 2 to 21. A drainage threshold of two square miles was used to develop stream networks for different region sizes. This threshold size was selected because floodplains modelled from a drainage threshold of two square miles have a higher degree of overlap with FEMA Q3 floodplains than floodplains modelled from other drainage thresholds. In contrast to the notably different stream networks at different drainage thresholds, the stream networks at different region sizes look identical, with variations that cannot be observed at the current mapping scale. We selected stream reaches intersecting Ozark to create the flood scenario. HAZUS hydrologic and hydraulic analyses as well as loss estimates were then conducted. The total floodplain area of Ozark was compared (for different region sizes) after the modelled floodplain was clipped to the city boundary. The estimated flood debris and economic loss were directly compared since the estimates are based on comparable stream reaches. We used the modelling results for a one-county study region as a benchmark, understanding that a large portion of hazard mitigation plans using HAZUS are prepared for local, county-level jurisdictions (FEMA 2005).

The floodplains modelled at different drainage thresholds are within a similar range of flood depth from 0.0001 to 6.74 m, while the total floodplain area (within Ozark city boundaries) has a decreasing trend with an increasing drainage threshold. The change rate is substantial, ranging from -20 to 51 percent when using the floodplain area at a drainage threshold of two square miles as a benchmark. However, at drainage thresholds of one square mile and larger, the variation decreases to within 20 percent. The decreasing trend is mainly due to the fast reduction of small tributaries when drainage threshold increases. The decreasing trend of stream length with increasing drainage threshold is consistent with findings of Stepinski and Collier (2004) and Moussa and Bocquillon (1996). Notice that the trend approximates a logarithmic function and become relatively level at a drainage threshold of two square miles, indicating that most reaches intersecting Ozark are within a drainage threshold of two square miles. The implication is that stream reach delineation for the city of Ozark will be sensitive to a specified drainage threshold within two square miles, while stream reaches delineated based on a drainage threshold of two square miles and larger will be similar.

Compared with FEMA Q3 floodplains, HAZUS underestimated the total floodplain area at all drainage thresholds; the largest floodplain coverage modelled at a drainage threshold of 0.95 square miles underestimated 2.71 percent of the total floodplain area. The under-estimation of floodplain areas by HAZUS is consistent with findings reported by Ding et al. (2008) and Gall et al. (2007). The degree of overlapping between HAZUS-modelled floodplains and FEMA Q3 floodplains was determined using producer's and user's summary accuracy statistics. The producer's accuracy is calculated by dividing the total overlapping area with the total area of FEMA Q3 floodplains, and the user's accuracy is calculated by dividing the total overlapping area with the total area of HAZUS-modelled floodplains. They are viewed from different perspectives: the producer's accuracy indicates the percentage of real floodplain areas which are correctly modelled, whereas the user's accuracy indicates the percentage of modelled floodplain areas that are correct.

We calculated the accuracy statistics for modelled floodplains within the boundaries of Ozark city and in its downtown areas where the concentrations of population and critical facilities are high. Accuracy assessment shows that floodplains modelled at a drainage threshold of 0.25 square miles have the highest producer's but the lowest user's accuracy for floodplains within the city limits of Ozark and its downtown area. This is because floodplains modelled at 0.25 drainage threshold cover most of the FEMA Q3 floodplains while many modelled tributary floodplains are spurious, particularly in south Ozark downtown areas. Floodplains modelled at a drainage threshold of two square miles had the second highest producer's accuracy as well as considerably high user's accuracy. Therefore, a drainage threshold of two square miles was regarded as optimal for floodplain modelling in the Ozark study area.

In contrast to the decreasing trends of the total floodplain area with increasing drainage threshold, the total amount of flood debris had an overall increasing trend, with a substantial change rate from -46 to 14 percent to the benchmark at the optimal drainage threshold of two square miles. The increasing trend of flood debris can be explained by a slower basin outflow and a longer flow time at a larger drainage threshold.

The total economic loss decreased with increasing drainage threshold, with a change rate from -20 to 24 percent to the benchmark at the optimal drainage threshold. The decreasing trend is mainly due to fewer floodplain areas as floodplains at small drainage thresholds not only cover the floodplains at large drainage thresholds but also include additional floodplains at small tributaries. It is worth noting that the impact of indirect economic loss on total economic loss becomes more important as tributary floodplains increase with increasing drainage threshold, the reason being that properties on tributary floodplains have higher economic value than do properties in the floodplains of main streams because of higher population density and business activity. The total floodplain area does not have a clear increasing or decreasing trend with changing region size, though the total floodplain area tends to be larger at a larger region size.

The total floodplain has a change rate ranging from -23.6 to 0 percent compared with that of the benchmark at a one-county study region. For a region size from 0.4 to 2.2 times a one-county area, the variation is about 5.3 percent. The total reach length fluctuates to within 7.6 percent compared with that of the benchmark as the region size increases. The small change rate is due the small change of stream networks at different region sizes. Unlike the correspondence found between reach length and floodplain area when drainage threshold changes, no corresponding changes of reach length and floodplain area were found with varying region size.

Compared to the benchmark, the total amount of flood debris fluctuated from -42 to 30 percent with changing region size. Similarly, the total economic loss did not have a consistent trend, fluctuating between -37 to 9 percent compared to the benchmark. For a region size ranging from 0.4 to 2.2 times a one-county area, the total economic loss had a small variation within 13 percent compared to the benchmark. This small variation corresponded to the small variation of floodplain area at large region sizes. This correspondence is consistent with our previous findings in the analysis of drainage threshold effects; the total floodplain area appears to be a major determinant of the total economic loss.

This study investigates the effects of drainage threshold and region size on HAZUS flood modelling. While the derived error bounds of flood estimates are only applicable to the Ozark study area, the trends of flood estimates with changing drainage threshold are likely similar for other geographic areas. This study determined the optimal drainage threshold to be two square miles by the degree of overlapping of modelled floodplains with FEMA Q3 floodplains. It would be interesting to investigate whether the optimal drainage thresholds for other study areas are the same using the proposed approach.

As the trends of estimated floodplain area and economic loss become stable at a region size of 0.42 local-county-areas and larger, a study region comprising the local county appears to be a logical, as well as convenient, choice for future HAZUS flood modelling. Stream reaches moving across the interested areas can be subsequently selected to conduct flood analysis. The study showed that HAZUS flooding modelling can be sensitive to changing drainage threshold and region size. The optimal drainage threshold was found to be two sqaure miles, while the optimal region size was determined to be the local county. The corresponding trends found for total floodplain area and economic loss with changing drainage threshold and region size change indicate that the total floodplain area is a major determinant of the total economic loss. The results of this study allow flood researchers of the Ozark area to understand the reliability of HAZUS flood loss estimates. The findings of this study also provide insights for other study areas to determine appropriate HAZUS parameters. Importantly, this study has established the methodology to perform future investigation of the effects of drainage threshold and region size on HAZUS flood modelling.

# Chapter 3

## Rationality, Spirituality and Shared Waters

Water management, by definition, is conflict management. Water, unlike other scarce, consumable resources, is used to fuel all facets of society, from biologies to economies to aesthetics to spiritual practice. Moreover, it fluctuates wildly in space and time, its management is usually fragmented and it is often subject to vague, arcane and/or contradictory legal principles. As such, there is no such thing as managing water for a single purpose all water management is multi-objective and based on navigating competing interests. Within a nation, these interests include domestic users, agriculturalists, hydropower generators, recreators and environmentalists. Any two of the interests are regularly at odds, and the complexity of finding mutually acceptable solutions increases exponentially as more stakeholders are involved. Add international boundaries, and the difficulty grows substantially yet again.

While press reports of international waters often focus on conflict, there are encouraging stories throughout the world on how water also induces cooperation, even in particularly hostile basins, and even as disputes rage over other issues. This has been true from the Jordan Basin (between Arabs and Israelis) to the Indus Basin (between Indians and Pakistanis) to the Kura-Araks Basin (among Georgians, Armenians and Azeris). Despite empirical research that repeatedly shows how water-related cooperation has vastly exceeded conflict over the last fifty years, prevailing theories fail to explain this phenomenon. Certainly, there is a long history of conflicts over, or related to, shared freshwater resources; yet, there is also a long and in many ways deeper history of water-related cooperation. Why do countries that share a basin cooperate on water, even when they will not cooperate over other issues? Water is a resource upon which we are all dependent and for which there is little detailed guidance in international law. By any quantitative measure, water should be the most conflictive of resources, not an elixir that drives enemies to craft functioning and resilient institutional arrangements.

Studies offer economic, environmental or strategic rationale to explain this "hydro-cooperation," but none of these studies seem completely adequate. Prevailing wisdom in both the science and policy of water resources does not seem to provide the foundation for answering this clearly ethical question. Perhaps some part of the answer lies not in the world of rationality, but rather in the spiritual, ethical and moral dimensions of water conflict resolution. Incorporating these components may offer not only new understanding of current disputes, but also models, tools and strategies for more effective water conflict management and transformation in the future.

This chapter seeks to investigate the potential of integrating a spiritual understanding of water conflict transformation with currently prevailing economic, environmental and strategic constructs. First, the context of the current understanding of water conflict and cooperation is presented. Then, the geography of what I call the Enlightenment rift the process by which the North/West separated out the worlds of rationality from spirituality is investigated by exploring the impact this rift has on ideas related to natural resource management. This idea is then developed under the context of the current clash of worldviews, as the North/West entwines its rational construct with the flow of international development capital and management philosophies, and the inevitable disconnect as these approaches collide with the more integrated views of the South/East. In closing, this chapter describes how the two worldviews might be gently interwoven within a fairly universal construct of the Four Worlds of perception, and how this construct might be employed within the framework of more effective water conflict management and transformation.

## Water, Conflict and Cooperation

Water is a unique and vital resource for which there is no substitute. It ignores political boundaries, fluctuates in both space and time and has multiple and conflicting demands on its use problems compounded in the international realm by the fact that the international law that governs it is poorly developed, contradictory and unenforceable. It is no wonder, then, that water is perpetually named not only as a cause of historic armed conflict, but also as the resource that will bring combatants to the battlefield in the 21st century What is the likelihood that "the wars of the next century will be over water," as some have predicted?

In order to cut through the prevailing anecdotal approach to the history of water conflicts, researchers at Oregon State University (OSU) undertook a three year research project, which attempted to compile a dataset of every reported interaction between two or more nations, whether conflictive or cooperative, that involved water as a scarce and/or consumable resource or as a quantity to be managed, i.e., cases in which water was the driver of events. The study documented a total of 1,831 interactions, both conflictive and cooperative, between two or more nations over water during the past fifty years. The results challenge the notion that water will lead to the next big wars of our time.

First, despite the potential for dispute in international basins, the record of acute conflict over international water resources is historically overwhelmed by the record of cooperation. The last fifty years have seen only thirty-seven acute disputes, i.e. those involving violence. Among these disputes, thirty took place between Israel and one of its neighbours, and that violence ended in 1970. Non-Middle East cases accounted for only five acute events, while, during the same period, 157 treaties were negotiated and signed. In fact, the only water war between nations on record occurred over 4,500 years ago between the city-states of Lagash and Umma in the Tigris-Euphrates basin. The total number of water-related events between nations of any magnitude is likewise weighted towards cooperation: 507 conflict-related events versus 1,228 cooperative events. This implies that violence over water is strategically irrational, hydrographically ineffective and not economically viable.

Second, despite the occasional fiery rhetoric of politicians perhaps aimed more often at their own constituencies than at the enemy most actions that occur over water are mild. Of all the events, some 43 percent fell between mild verbal support and mild verbal hostility. If official verbal support and official verbal hostility are included in the analysis, the share of verbal events reaches 62 percent of the total. Thus, almost two-thirds of all events were only verbal, and more than two-thirds of those had no official sanction.

Third, cooperative events related to water covered a broad spectrum, including water quantity, quality, economic development, hydropower and joint management. In contrast, almost 90 percent of the conflict-laden events related to quantity and infrastructure.

Furthermore, almost all extensive military acts the most extreme cases of conflict fell within these two categories. As such, water acted as both an irritant and a unifier. As an irritant, water can make good relations bad and bad relations worse. Despite the complexity, however, international waters can act as a unifier in basins with relatively strong institutions.

This historical record suggests that international water disputes do get resolved, even among enemies, and even as conflicts erupt over other issues. Some of the world's most vociferous enemies have negotiated water agreements or are currently in the process of doing so. The institutions they have created often prove to be resilient, even when relations are strained. For example, the Mekong Committee-established by the governments of Cambodia, Laos, Thailand and Vietnam as an intergovernmental agency in 1957 exchanged data and information on water resources development throughout the Vietnam War. Israel and Jordan have held secret "picnic table" talks on managing the Jordan River since the unsuccessful Johnston negotiations of 1953-1955, even though they were technically at war from Israel's independence in 1948 until the 1994 treaty The Indus River Commission survived two major wars between India and Pakistan. Moreover, all ten Nile basin riparian countries are currently involved in senior government-level negotiations to develop the basin cooperatively, despite water wars rhetoric between upstream and downstream states.

Therefore, the question of whether water leads to conflict or cooperation turns out to not be mutually exclusive at all both are true, and often in the same place. Shared water often leads to tensions between nations, which in turn offer a vehicle for dialogue that often results in some form of joint management. In fact, a general pattern has emerged for international basins over time that exemplifies both conflict and cooperation. Riparians of an international basin implement water development projects unilaterally first on water within their territory, in attempts to avoid the political intricacies of the shared resource. At some point, one of the riparians, generally the regional power, will implement a project that impacts at least one of its neighbours. These projects can, in the absence of relations or institutions conducive to conflict resolution, become a flashpoint, heightening tensions and regional instability that, in turn, lead to negotiations, treaties and/or collaborative river basin organisations and mechanisms for future conflict management.

These processes from conflict to cooperation have historically suggested room for greater efficiencies. The first complicating factor is the time lag between when nations first start to impinge on each other's water planning and when agreements are finally, arduously, reached. As noted, once a project becomes a flashpoint that heightens tensions and regional instability, it often takes years, or more commonly, decades, to resolve. The negotiations for the Indus, Ganges and Jordan treaties took about ten, thirty and forty years, respectively All the while, water quality and quantity degrades to the point at which the health of dependent populations and ecosystems are damaged or destroyed. This problem gets worse as the dispute rises in intensity; one rarely hears talk about the ecosystems of the lower Nile, the lower Jordan or the tributaries of the Aral Sea because they have effectively been written off to the vagaries of human intractability. During these periods, threats and disputes rage across boundaries with relations as diverse as those between Indians and Pakistanis and between Americans and Canadians. Water was the last and most contentious issue resolved in negotiations over a 1994 peace treaty between Israel and Jordan, and was relegated to "final status" negotiations along with some of the other most difficult issues such as Jerusalem and refugees between Israel and the Palestinians.

The second set of security issues occur at the subnational level. If there is a history of water-related violence and there is correction internal water conflicts, ranging from interstate violence and death along the Cauvery River in India. The inland, desert state of Arizona even commissioned a navy consisting of one ferryboat and sent its state militia to stop a dam and diversion on the Colorado River in 1934. In fact, the recent research at OSU suggests that, as the scale drops, the likelihood and intensity of violence rises. This leads to the realisation among the water community that despite the historic record of cooperation over shared waters, there is still definite room for improvement.

## Reconnecting Process with Spirit

How can the process from conflict to cooperation be enhanced? To begin our understanding, we might drop our scale of analysis from the macro to the micro level. Along with describing global and regional trends from an abstract geopolitical perspective, there is also the process that occurs "in the room." At the end of the day, negotiations are about people and relationships, not solely about geopolitics and economic interests. This begs the question: Are negotiations rational or is something else taking place "in the room," something connected more to energy and transformation? To gain insight into this question, it is worth looking at the values and philosophies inherent within the negotiating context. This approach will be developed in the following sections.

### *The Enlightenment Rift*

The Enlightenment of the 18th century left a profound rift between the worlds of spirit and reason, one with intense implications for today's clash of ideas. In temporal terms, it suggested that day-to-day considerations should be gauged in rational, objective concepts, while the world's spiritual dimension should be considered separately, in the evening at home or within one's Friday, Saturday or Sunday community. Over time, rationality dictated the structure of subsequent paradigms, such as economics, science and modernity, to the point that those in the North and West are consistently satisfied to ask the "what" without the "why" in terms of public discourse. We talk comfortably of economic growth rates, for example, without the accompanying discussion of how focusing on creating and owning more commodities might affect an individual's soul. We are able to put the emphasis in debates over crime disproportionately on the value of punishment and retribution, and less regularly on the potential for the individual and his or her community for rehabilitation. We frequently turn to benefit-cost analyses as a decisionmaking tool in which all factors must be reduced to an economic value, explicitly excluding often profound, but intangible, considerations.

However, the idea of separating out rationality from spirituality is, fundamentally, a Western construct. As Smith eloquently puts it, "The modern West is the first society to view the physical world as a closed system." Conversely, much of the thinking in the South and East often retains its integration of rationality and spirituality.

As a model, which shows frameworks from two different spiritual traditions that illustrate the idealised relationship between self and community, between justice and mercy, and between boundaries and expanse? This illustration models three of the ten Kabbalistic sefirot, or spheres of divine attributes. These three figures show a balance between din the attribute of justice, boundaries and self with the sefirah of chessed the attribute of loving-kindness and mercy. Within this tradition, these two attributes remain in balance one cannot exist without the other, but not quite. In this balance of divine attributes, the sefirah of chessed, is always modelled just a touch higher, connoting that attribute takes precedence in any conflict between the two.

As the map of the sefirot shows, the balance of din (justice) with chessed (mercy) is manifested in the sefirah of rachamim (compassion). The attribute of rachamim is explicit in what is meant by compassion, suggesting a precise integration of consideration for both justice and mercy, for self and community, for boundaries and expanse. The root of racharnim is the same as for rechern (womb) giving us a very clear allusion to what is meant: A mother is able to give not only nourishment, but her very life force to her unborn child (ultimate loving-kindness), and is able to do so only if she takes care of her own health and needs.

The message of the construct has relevance for many of today's grand issues. Do we pursue justice or mercy in our lives and politics? Shall we be concerned with individual rights or responsibility to one's community? Modernity or post-modernity? Free market or safety nets? Right or left? Red or blue'? Given the understanding elaborated above, the answer to all these choices is "yes, in exquisite balance." The dichotomies are false, as is the apparent division between rationality and spirituality.

This balanced construct exists fairly universally and can be influential in respect to approaches to resource allocation, negotiations and understanding of relationships. The unity of a balance of self and other, light and dark, can be seen in the Taijitu, the traditional Taoist symbol for Yin and Yang. In the Christian construct, the triad of justice, loving-kindness and compassion has been described through the Trinity the Father, Son and Holy Spirit. Some Christian Kabbalists make these comparisons explicit. In Islam, Al-Hakam (the Judge), Ar-Rahman (the Merciful) and Ar-Rahim (the Compassionate) are three common names out of the ninety-nine names of Allah.

To generalise, the overemphasis on rationality and the rights of the individual as opposed to the inclusion of spirit and the needs of the community is disproportionately a North/West phenomenon, associated primarily with the non-Asian developed world. The South/East often retains a more integrated view of issues of the individual with the community or one's spirituality with one's rationality. These are two profoundly contradictory worldviews: The North/West's dichotomous views of rationality and spirituality, justice and mercy, are in stark contrast to the South/East's holistic, integrated balance. They clash regularly and intensely across the world stage, from foreign policies to expectations of immigrant communities to dynamics in the United Nations. In other fora, one might note the implication this geography has on the current clash of civilizations, but one can use water as a microcosm of these larger issues.

## The Enlightenment Rift and Water Ethics

### *Water and the Economics of Cooperation*

The geography of this post-Enlightenment rift is, at times, enlightening. What this figure illustrates is the extensive interface between very different value structures. In recent decades, for example, the North/West has approached international water management from an increasingly economic framework, most notably through the 1992 Dublin Principles that state: "Water has an economic value in all its competing uses and should be recognised as an economic good." This was the first explicit recognition of water as an economic good, and this principle is often found quoted in literature that has ensued since its establishment. Agenda 21, which emanated from the Rio Conference on Environment and Development in June of 1992, echoed this theme. Furthermore, the World Bank and other development banks have increasingly been urging conflict resolution by thinking of water less in terms of a zero-sum commodity and more in terms of the benefits of water, a positive-sum commodity that can be enhanced and quantified through economic principles.

Yet, these economic principles, so prevalent in the North/West and encouraged through its development agencies and banks, explicitly contradict local and indigenous practices throughout the developing world. For example, Islamic legal tenets will vary in accordance to the source of the water. Water "provided by God," such as natural surface or groundwater available year-round, will have different legal implications than water "provided by man" such as a cistern or the attendant canal system requiring human labour. "God-given" waters may not be bought or sold, and their use is available to all equally. To many, the idea of buying and selling water is both repugnant, like "buying and selling one's children," as one Berber mediator mentioned in an interview, and contrary to the tenets of Islam.

### *Healing Waters: Water Rights and Water Honour*

Another interface for these clashing concepts and principles is in the realm of conflict management and transformation. As noted above, approaches to the balance between the individual and the community can be thought of very differently in the North/West as compared to the South/East. These views play out in approaches to legal principles and conflict resolution as well. The Western-based legal structure, for example, is very comfortable with the idea that one side in a dispute can be found entirely right and gain everything, while another side can be found entirely wrong and lose everything. In contrast, in many communities, the balance of justice and mercy described above leads to the importance of both retaining individual rights and honour on both sides of a dispute, and to reconciliation of a wrongdoer within one's community. For example, consider the Arabic word taarradhin defined as, "resolution of a conflict that involves no humiliation." It is a profound concept with no Western equivalent. This is not a theoretical concept, but a concept put into practice throughout the Muslim world when a wrong has been committed. This takes place through the ceremony of a sulha, a ritual ceremony of forgiveness. The term comes from musalaha (reconciliation), which implies that hostilities are ended, honour re-established and peace restored in the community. This custom, which consists of private, often mediated, negotiation of redress between the affected parties, is followed by a public declaration of forgiveness and usually a festive meal. Once the sulha is performed, the slate is wiped clean, as if the dispute never happened. The agreement is legally binding on both the individuals and the community. Grudges are dissuaded and reference to past disputes may not be made to gain position in a current conflict.

The balance of rights and honour, and justice and mercy, in contrast to the Western construct of justice, is described by Jabbour:

This is how [social] justice should be achieved. The courts condemn the guilty party in vain, because they never take care of the harm. Magistrates and police don't know what social justice is. Honour is an alien virtue. They believe in the virtue of punishment, but forgiveness is overshadowed and neglected, because peacemaking is not on their minds.

The international community seems to be lacking in this very type of ritual ceremony of forgiveness. The negotiating process of many transboundary agreements is secret or, at best, a televised signing ceremony. Accord over an issue such as water, generally considered unnewsworthy, may take place without any public notice at all. A public ceremony would allow the community affected by a dispute the stakeholders from all sides to celebrate its resolution and thereby assume ownership of its implementation. To be fair, the field now known as Alternative Dispute Resolution (ADR) came about precisely because of the limitations, and cost, of the Western legal framework. For the most part, even ADR, as exemplified by Fisher and Ury and its offshoots, is more often than not rooted in both rationality and/or economic constructs of interests and benefits. A claim by practitioners is that ADR works because the mediator can help parties construct agreements that meet the needs of both parties. In other words, people agree when it is in their interest to agree. But how does one know it was in the parties' best interests to agree? The only proof is tautological: they agreed.

## Spiritual Transformation and Lessons for Water Negotiations

### *Conflict, Cooperation and Kabbalah*

Are negotiations actually rational? Do we agree when it is in our interest to agree? Rather, is something more transcendent occurring in the negotiating room a process more connected to energy, and transformation? Successful multiparty negotiations often require profound transformations in the way participants conceptualize the issues at hand. Those involved can often define the precise moment when thinking altered dramatically, essentially the "aha!" moment where emphasis shifted from individuals thinking solely in terms of their own respective agenda to a dual comprehension and understanding of others' needs. As previously noted, traditional conflict resolution models define these transitional moments in rational terms: People come to agreement when it is in their interest to agree. Even overlooking the tautological nature of this argument, rationality simply often does not hold sway if the conflict involves even a modicum of real emotion.

To really understand the process of transformation, and the settings most conducive to inducing these shifts, one should look outside of the field of conflict resolution as currently defined in modern, academic terms. When one thinks of the situations most analogous to settings conducive to transformative thinking, the world of spiritual transformation arises as perhaps the most appropriate. After all, every spiritual tradition in the world is devoted to precisely this process of transformation: to transform and aid individuals in moving from a focus on their own immediate wants and desires, to an acknowledgment that focuses on addressing their obligations to society, humanity and the divine. Throughout the history of water conflict management, inducements to diverse stakeholders have centred on three sets of interests: economic inducements, environmental protection and strategic interests. Each has met with successes over the years, yet each has its limitations, especially in tense settings. Researchers have occasionally sought to bolster these traditional approaches with the experiences of other, seemingly peripheral, issues. For example, a report on a workshop that brought together participants involved in international water management and international arms control agreements suggests that common issues were found, and approaches exchanged, in regards to political suspicion, incomplete data and the monitoring and enforcement of agreements. Ostrom has undergone remarkable work in tying small-scale, local experiences in water management with larger lessons and scales, and Wolf investigates the allocation rules and conflict resolution mechanisms of Berbers and Bedouin and draws implications from their experiences for international waters.

Most research and practice of connecting the worlds of spirit and water has focused on how religions address the environment and its protection. Much of this thinking has been documented by Mary Evelyn Tucker and John Grim of Bucknell University, who coordinated a ten-conference series on World Religions and Ecology at Harvard's Centre for the Study of World Religions, and who are editing the conference papers from the series, with separate volumes describing relationships between each religion and the environment. Similarly, Palmer and Finlay provide an overview of religious approaches that include statements on the topic from leaders in eleven of the world's religions, as specifically related to the environment's connection to international development, and include statements on the topic from leaders in eleven of the world's religions.

However, little in this field has been specifically applied to shared water and issues of environmental conflict management. Some work has described the Catholic Church's leadership in the promotion of a religious obligation to protect the Columbia River watershed as an international basin. Further, Ecumenical Patriarch Bartholomew recently planned a boat trip down the Danube with international invitees to promote protection of that basin, although the trip was subsequently cancelled.

The focus here is subtle, but importantly different from that described above. The key is what spiritual and ethical processes of transformation can offer environmental negotiations and conflict management, not the overall framework in which each religious tradition approaches the environment as a whole. As such, the scope of each individual application is by necessity defined on a smaller scale by a select group of stakeholders. Yet, one could argue that the overall potential to effect change in thinking about the process of conflict management is vast.

Very little work has been done explicitly in tying spiritual transformation to environmental negotiations and management, although what has been done suggests exciting potential. In February 2003, the Program on Negotiation at Harvard Law School launched the Harvard Negotiation Insight Initiative, specifically aimed to offer mediators and stakeholders resources to "explore the interface of contemporary negotiation theory and practice with alternative frameworks including some drawn from perennial wisdom traditions." While the initial focus was in the realm of legal mediation within the United States, the first workshop, held in June 2004, brought a variety of facilitators, mediators and environmental experts together for training. As the annual workshops have grown, the focus has broadened to encompass all branches of mediation and negotiation.

Literature does exist suggesting the potential for applying the tools of spiritual and ethical processes to conflict transformation possess, which goes on to further suggest the potential applications that these tools hold in relation to environmental issues as well. Transformative mediation, a relatively new branch of ADR, is based on "empowerment and mutual recognition of the parties involved," as well as on their long-term interests. It offers an alternative to problem-solving mediation, which can be highly directive and focused on short-term problem-solving. Compassionate listening is a faith-based technique of guided communication which has proved effective in extremely hostile settings, notably by Carol Hwoschinsky in guiding dialogue between Israelis and Palestinians who have been touched by violence. Similarly, Rabbi Uzi Weingarten teaches similar techniques of communicating with compassion, which have been applied in a variety of hostile settings. Abou El Fadl describes Islamic processes for "institutionalising mercy and compassion in social interaction," while McConnell structures mediation in a Buddhist construct, and Barthel suggests lessons for process from a Baha'i perspective.

The Association for Conflict Resolution (ACR) now has a spirituality section, and the Fall 2005 issue of ACR's journal, AC Resolution is entitled "Spirituality and the Heart of Conflict Resolution." While most of that activity focuses on the mediator's own spirituality, or on mediation as a spiritual practice, some work does draw directly from the spiritual world to facilitate the process of conflict resolution. Marc Gopin, director of the Centre for World Religions, Diplomacy and Conflict Resolution at George Mason University, suggests the potential for conflict resolution's learning from spiritual transformation:

One example [of the possibilities of merging study of religion and conflict resolution] is the spiritual process of transformation of character through reflection and ethical improvement of one's behaviour. Several theories of conflict resolution suggest the importance of personal transformation for the resolution of deep conflicts. Spiritual programs of personal transformation might be combined with this kind of conflict resolution methodology, in religious settings could such phenomena be incorporated into conflict resolution strategies among religious peoples or even more generally?

Process and the Four Worlds Thinking of negotiations less in terms of rational interests and more in terms of transforming energy allows us to centre on the process of transformation in negotiations the point at which parties move from thinking of themselves as representing countries or political bodies to perceiving more broadly the needs of all stakeholders and needs within a basin. These are critical junctures that occur within the negotiations processes where movement from rights-based to needs-based to interest-based to equity-based negotiations suddenly becomes possible. In international basins, as noted above, this transformation may normally take years or even decades, during which time political tensions are exacerbated, ecosystems go unprotected and water is inefficiently managed, at best. However, this negotiation transformation may have a corollary in spiritual transformation. Every spiritual tradition in the world is devoted to a very similar process, which includes guiding individuals to thinking less about their immediate wants and desires, and to encouraging more of a focus on their obligations to society, humanity and other issues larger than themselves. In this setting, conflict can be seen less as a displacement between rational sets of interests, and more as a rift in the fabric of community, with the attendant obligation for healing.

### *The Universality of the Four Worlds*

One construct that can help inform negotiation processes is the idea of the Four Worlds and the use of transformative processes to move through them. In many faith traditions, our relationship to the world can be experienced through four types of perception: physical, emotional, knowing and spiritual. One intuitive example might be seen through a glass of water, which exists most recognisably on a physical plane. However, if one is thirsty or the water is particularly satisfying, one's experience of water can be transformed into an emotional response. One can also intellectualize the water and thus can consider its components and interaction of the water with our body to provide and maintain sustenance. Finally, one might say a blessing over the water, lifting its "profane" covering, and it now becomes a source of spiritual nourishment. While these four levels of perception can be considered separately, they can often be best achieved in sequence but they should neither be considered distinct nor linear. The water exists simultaneously in all four states and it is up to the individual to determine through which lens it will be best perceived. Nonetheless, understanding the Four Worlds in sequence is often useful, if not critical.

This construct seems particularly useful for our purposes of cross-cultural and political boundary negotiations. One notes the near universality of the Four Worlds construct, which Huston Smith refers to as the "levels of reality." Psychologists will recognise Maslow's hierarchy of needs in the Four Worlds, but those familiar with the mystical traditions around the globe will find much more ancient roots. In Judaism, the Four Worlds are seen in Kabbalah, and applied through the structure of the daily prayer service a guided meditation based, in turn, on the construct of the Temple in Jerusalem, which itself was built to emulate Moses' experience of the levels of holiness during his ascent of Mount Sinai. In Buddhism, these levels are described in the Four Jhannas levels achieved through meditation and corresponding to "the four great levels of the heavenly realm." They are evident within the Sufi mystical experience, as well as in the Shaman Four Levels of Perception, the Toltec Four Agreements and the Native American Sacred Tree. In fact, this construct is so widespread that Zalman Shachter-Shlomi suggests that human biology is actually hard-wired to experience the world through these four lenses the reptilian, limbic, cortex and unused portions of the human brain.

Understanding this construct and its universality leads to tremendous possibilities in relation to the design and implementation of the negotiation processes, training and collaborative learning. It allows for a structure that moves through different lenses and perceptions, while tapping into what seems to be a fairly universal set of needs. Finally, it allows for a focus on the transformative processes in negotiations to bolster the only partially successful historical emphasis on quantifiable benefits.

### Application: Stages of Water Conflict Transformation

How might the Four Worlds be harnessed for water negotiations? This section describes how each of the worlds can be seen in different stages of water conflict and conflict transformation. There are, of course, no blueprints for water conflict transformation. Gilbert E White reminded us long ago that, "every basin is unique." There does seem to be, however, general patterns in approaches to water conflict which have emerged over time. Classic disputes between, for example, developers and environmentalists, rural and urban users, or upstream and downstream riparians, suggest zero-sum confrontations where one party's loss is another's gain and when confrontation seems inevitable. Yet such seemingly intractable conflicts are regularly and commonly resolved, as creative thinking and human ingenuity allow solutions which draw on a more intricate understanding of both water and conflict to come to the fore.

Over time, the process of transformation along a continuum from zero-sum, intractable disputes towards positive-sum, creative solutions has been slightly formalised and can be described through a migration generally along four stages. Note that all stages exist simultaneously, and therefore, do not need to be approached in a sequence nor achieved necessarily for success. In today's world, many disputes never move beyond the first or second stage, yet are tremendously resilient, while a few have achieved the fourth stage and are fraught with tension. Nevertheless, like any skill, it is useful to understand the structure of an ideal path in order to perfect the tools required for any individual situation. The generalised path described here is structured around an understanding of each of the four stages through any of four perspectives.

### Stage I. Assess the Current Setting: Basins with Boundaries

In Stage I, in its initial, adversarial setting, regional geopolitics often overwhelms the capacity for efficient water resources management. Metaphorically, the political boundaries on a map at this stage are more prevalent than any other boundaries, either of interest, sector or hydrology. Dialogue is often focused on the past, based on the rights to which a country, state or province feels it is entitled, and a period of expressing of pent-up grievances can be necessary As a consequence of these initial tensions, the collaborative learning emphasis is on trust-building, notably on active and transformative listening, and on the process of conflict transformation. By focusing primarily on the rights and interests of countries, states and/or provinces, inefficiencies and inequities are inevitable. Once stakeholders are brought to the table, this stage generally involves classic hydropolitical assessments of the current setting within a basin, including biophysical, socioeconomic and geopolitical parameters. The processes for assessing many of these aspects are well-defined (e.g., hydrologic studies or benefit-cost analyses of development alternatives), while many are less quantitative, but no less critical (e.g., social impact statements or assessments of indigenous traditions of management).

At this stage, stakeholders often think nationally, or as a state or province or other political constituency, and are focused on their rights. This may lead them to look disproportionately backwards, if only to be able to vent and perhaps address perceived grievances. Although understanding the baseline of any basin may take decades (if possible at all), it is not necessary to agree to all data before greater cooperation takes place. These assessments or training workshops can be used in and of themselves as confidence-building measures to move to the next stage, even as greater mutual understanding of the basin is being created.

### *Stage II. Changing Perceptions: Basins without Boundaries*

As the adversarial stage plays out, occasionally some cracks can be seen in the strict rights-based, country or state/province-based positions of each side. In actual water negotiations, this process can last decades. Eventually, and sometimes painfully, a shift can start to take place where the parties begin to listen a bit more, and where the interests underlying the positions start to become a bit apparent. In Stage II, a reflexive stage, negotiations can shift from rights (what a country, state/province feels it deserves) to needs (what is actually required to fulfill its goals). Conceptually, it is as if we have taken the national, state/provincial boundaries off the map and can, as if for the first time, start to assess the needs of the watershed as a whole. This shift, from speaking to listening, from rights to needs and from a basin with boundaries to one without, is a huge and crucial conceptual shift on the part of the participants, that can be both profoundly difficult to accomplish and absolutely vital to achieve for any movement at all toward sustainable basin management. To help accomplish this shift, the collaborative learning emphasis is on skills-building, and we might approach the "boundary-less" basin by sector rather than by nation.

At this stage, the attention shifts from past to future, as stakeholders examine each others' interests beyond positions. A process of social learning sets in. Parties can begin to ask, "What could be?" rather than "What was?" or "What is?" The metaphor for this stage is a basin without borders where, rather than rights, there are needs; and rather than thinking of national issues, we might look instead to how different sectors might be developed basin-wide.

This shift is transformative the point at which parties move from thinking of themselves as representing countries or states/provinces to perceiving more broadly the needs of all stakeholders within a basin (whether or not they like these needs). Parties begin to understand the needs of the other and thus the requirements that must be met if agreements are to be reached.

### *Stage III. Enhancing Relations and Benefits: Beyond the River*

Once participants have moved through the first two stages from mostly speaking to mostly listening, and from thinking about rights to thinking about needs, the problem-solving capabilities that are inherent to most groups can begin to foster creative, cooperative solutions. In Stage III, an integrative stage, the needs expressed earlier begin to coalesce to form group interests the why underlying the desire for the resource. Conceptually, we start to add benefits to the still-boundary-less map, and in fact, to think about how to enhance benefits throughout the region, often by adding resources other than water, and geographic units other than the basin. In fact, rather than allocating water, we can think about allocating benefits. The collaborative learning emphasis is now on the relationship-building of the group, and we begin to move in "benefit-shed" rather than being restricted by the basin boundaries.

Once the shift has been made in thinking about allocating water to allocating benefits, it is a natural progression to think together about how to enhance the benefits within and beyond the basin. This may be done within the realm of water resources alone. A well-designed dam upstream might, for example, both enhance agricultural production downstream and help protect riparian habitat. But it is often helpful to think at this stage about "baskets of benefits" that may go well beyond water, or well beyond the basin in question. Indeed, the most successful cases of building regional approaches to water have gone beyond seeing water as the ends to seeing it as a means to achieve other goals, such as socioeconomic development and reduction of fears of floods and drought. Energy production and water development are often linked, for example, as are afforestation programs, transportation networks and environmental protection. Naturally, the transaction costs of including more sectors than water go up exponentially, but so do the potential benefits. This means bringing in actors beyond the water sector and expanding the basket to be considered.

### *Stage IV. Bringing it all Together: Institutional and Organisational Capacity and Sharing Benefits*

Finally, although tremendous progress has been made within the first three stages, both in terms of group dynamics and in developing cooperative benefits, Stage IV (the action stage) helps with tools to guide the sustainable implementation of the plans and to ensure that the benefits are distributed equitably among the parties. The scale at this stage is now regional where, conceptually, we need to put the political boundaries back on the map, reintroducing the political interest in seeing that the baskets that have been developed are to the benefit of all. The collaborative learning emphasis is on capacity-building, primarily of institutions.

Much as the water community likes to think in terms of basins or watersheds alone, eventually the borders have to come back on the map. Political entities are primarily responsible for their own benefits and sovereignty, after all, and it is often hard to sell their own constituents on an integrated basin alone. The most critical issues at this stage are addressing how benefits can be distributed equitably or perceived as fair; how sustainable and resilient institutions can be crafted; and how the existing institutions and organisations are to be taken care of or compensated for any change. The first question may require trade or side-payments, while the second and third questions must evoke the best in institutional design. It is important to remember that conflict potential can actually increase during periods or situations of increased benefits. The increase of benefits alone will not assure the mitigation of conflict. This is because parties may realise benefits they never had, but they may perceive that the other party is getting relatively more benefits than they are getting. Thus, the perceptions of fairness and equity, not just the tangible delivery of benefits, are critical. It is critical not to think of these stages as a linear process, where the further along the better. Most basins ebb and flow back and forth over time, finding the level that meets a particular set of hydropolitical needs for a given place and time there is no right set of answers. One might think of these all existing in parallel universes simultaneously, each with its own set of approaches or tools, any of which may be useful at any given time, or conceptually, as a helix or set of spheres rather than strictly linear. They are broken apart here only for the purposes of explanation.

## Four Worlds of Water

Understanding this construct may help structure more effective future negotiation processes, as well as skills-building and collaborative learning exercises. Even the word "water" can be understood differently depending on through which lens one is viewing it, and the mediator/facilitator can harness the construct and sequence of the Four Worlds to facilitate new understanding. In contrast, we ignore the Four Worlds at our peril.

As peace negotiations between Israelis and Arabs commenced in the early 1990s, for example, each side approached the issue of water very differently. From the Palestinian and Jordanian side, water was conceptualised in a very physical sense in some cases, people literally did not have enough clean water for sustenance or as "emotional water" the lack of sovereign control over water as representing larger issues of national identity and occupation. From the Israeli side, water was constructed intellectually, Survival had long been assured, so the challenge was to move, price, treat and store water in the most efficient manner.

These conflicting conceptualisations led not only to difficult impasses water was the last issue concluded in the Israel-Jordan Treaty of Peace but also to especially creative solutions. In what will no doubt become a classic modification of the tenets of international law, Israelis and Jordanians invented legal terminology to suit particularly local requirements ill their 1994 peace treaty. In negotiations leading up to the treaty, Israelis, making the intellectual argument that the entire region was running out of water, insisted on discussing only water allocations, that is, the future needs of each riparian. Jordanians, in contrast, refused to discuss the future until past grievances had been addressed. They would not negotiate allocations until the historic physical and emotional question of water rights had been resolved. There is little room to bargain between the past and the future, between rights and allocations. Negotiations reached an impasse until one of the mediators suggested the term rightful allocations to describe simultaneously historic claims and future goals for cooperative projects. This new term is now immortalised in the water-related clauses of the Israel-Jordan Treaty of Peace.

In relation to the Fourth World, we can see that, throughout the world, native and indigenous peoples see water as a holistic, spiritual resource. With the construct of the Four Worlds, we can conceptualize how jarring it can be, to the point of being sacrilegious, to approach problem-solving in rational, economic concepts. Shared water resources provide a useful lens through which to describe both the hazards of ignoring the relationship between rationality and spirituality, but also to demonstrate the potential an integrated approach may offer for effective negotiations and conflict transformation.

As the historically contrasting worldviews of the Global North/ West and Global South/East increasingly interact,, both within and without the worlds of shared waters, we have the opportunity to heal historic divisions. The history of water conflicts and cooperation suggest that people do come together, even across vociferous divides. It is anticipated, however, that the dangers of scarcity-driven suffering and conflict will only increase with population growth, poverty and global change. Yet, as grows the threat to increased conflict in the future, so too grow the opportunities for dialogue and healing. In 1996, the Episcopal Diocese of Massachusetts shifted its diocese boundaries from political divisions to watershed boundaries. The rationale was instructive:

Simply demonstrating that we are all connected by water: rich and poor, urban and rural, upstream and downstream, is a fine place to start. I think the Holy Spirit will take care of the rest. Water ignores all separations and boundaries save for those of the watershed itself. As such, it offers a vehicle to bring those who share it together. Since it touches all we do and experience, water creates a language through which we may discuss our common future.

# Chapter 4

## Water Resources Management

Transboundary water management poses difficult institutional and cultural challenges for U.S.-Mexico watershed initiatives, yet current and past experiences can offer useful lessons for other transboundary water managers and policymakers attempting to use coordinated resource management to better address water allotment and quality issues. While acknowledging that transboundary watershed initiatives in this region are relatively new and "their operational scope, mode of decision-making, and linkages amongst participating actors vary considerably by area and project throughout the border region" (Mumme 2002: 6), the potential for transboundary watershed initiatives to build on the experiences of others points to the value of a comparative case study. At the same time, generalizing about organizational structure and process of these initiatives is much easier than evaluating performance and productivity. Effective governance structures generally reflect specific local or regional contexts, yet if we base our discussion on the premise that "building local initiatives can advance regional cooperation on water resources" (Brown 2000), we must ask, What makes local initiatives successful? While it is too early in the development of U.S.-Mexico transboundary watershed initiatives to evaluate outcomes comprehensively, we can draw from the literature on watershed initiatives in the western United States and Mexico the following analytical criteria:

1. historical setting and current water resource issues;
2. organizational constraints, including clear identification of mission and focus with long-range vision and goals, monitoring and assessment, well-defined process rules, decision-making style, strong leadership, and consistent funding;
3. representation of all interests: state and municipal agencies, nongovernmental organizations (NGOs), and private

stakeholders, with the development of trust and mutual understanding; and

4. linkages between science, policy, and local stakeholder interests, and the education of participants and public in science and policy.

This case study examines two watershed initiatives in the upper San Pedro River basin, located in northeastern Sonora, Mexico, and southeastern Arizona. The basin was selected on the basis of our collective multiyear experiences working as participant-observers, policy analysts, program managers, and community collaborators. We begin with a brief account of the methodologies used in the case study and descriptions of the two watershed initiatives, stressing their similarities and differences. We then examine both initiatives in terms of (a) historical setting and current water resource issues, (b) organizational structures and processes, (c) representation of stakeholder interests, and (d) linkages between policy and science with basin stakeholders.

We also compare their relative successes and problems. We anticipate that the scale of the basin and specifically, the level of government at which resource-management problems are addressed and policy implemented will be important in terms of effective management. We also expect national governments to have a stronger role in the policy debate than will the states, despite decentralization playing an increasingly important role in the rise of regional watershed groups. We conclude with a look at the implications of these differences in the two portions of the basin for coordinated binational resource management and with a discussion of the significance of this case study for potential transboundary collaboration.

Surveys are frequently used to evaluate the performance and effectiveness of western U.S. watersheds, as are case studies involving interviews, text analysis, participant observation, or some combination thereof. Our approach was to interview agency members, NGO staff, municipal officials, water managers, legislators, and other policymakers; to use our own experiences as participant-observers, policy analysts, and collaborators; and to review case studies of transboundary watershed initiatives. The San Pedro River originates in northern Sonora, Mexico, and flows north into Arizona, eventually joining the Gila River, which flows into the Colorado River and later drains into the Gulf of California. The upper San Pedro River basin (USPB), which lies entirely within the Basin and Range Province, consists of the northwest-trending San Pedro River valley and the surrounding mountains, ranging from 4,200 feet (1,280 meters) to 3,300 feet (1,006 meters).The basin represents a transitional area between the Sonoran and Chihuahuan Deserts, with topography, climate, and vegetation varying considerably across the watershed.

The USPB, an area of approximately 1,875 square miles (6,400 kilometres), can be characterized as a mixture of desert and grasslands ecosystems with semiarid climate. Precipitation varies from a little more than 450 mm/year in Cananea, Sonora, at the southern end of the basin, to about 270 mm/year in the lowlands outside of Sierra Vista, Arizona. A change in land cover from grasslands to mesquite from 1973 to 1986 was the result primarily of climate fluctuations, livestock grazing, and more recently, rapid urbanization.

Approximately 114,000 people live and work in seven incorporated towns and several unincorporated communities in the two countries within the USPB. Population in the Mexican portion of the USPB is concentrated in Cananea and Naco, Sonora. Most of Cananea's 32,000 residents depend economically on the copper-mining operation that has been there for more than one hundred years. Closer to the border, Naco has approximately 5,300 residents, which can grow to 7,000, counting transient workers waiting to cross into the United States (INEGI 2000). Approximately nine ejidos, or communal agricultural settlements, are dispersed across the Mexican portion of the region. In the U.S. part of the basin, population is concentrated in Sierra Vista, with 38,000 residents, drawn largely from the army base at Ft. Huachucha and retirees.

Conflict over water issues needs to be understood in terms not only of scarcity or quality, but also in the context of conflicting attitudes and meanings developed over time; the local historical setting strongly influences international dynamics (Wolf 2002: 11). In the U.S. side of the basin, the legacy of the frontier Ft. Huachucha and an "independent spirit" characterize many of the stakeholders' reasons for moving to the area.

This perception has affected views of resource use, transformation of the environment, and private property rights and helps explain the pro-development position of some stakeholders. In the Mexican portion, Cananea has a reputation as the town that sparked the Mexican Revolution with its 1906 strike against the mine owner, William Cornell Greene, who once wanted to make Cananea a part of the United States. Cananea has had a century-long tradition of social activism that has included concern with environmental problems affecting community health, but the mine management's economic power and influence in the community have traditionally subdued environmental protest.

More recently, in 1998, the Commission for Environmental Cooperation (CEC) initiated a study of the San Pedro Riparian National Conservation Area (SPRNCA) characterizing the physical and biological conditions required to sustain and enhance the riparian migratory bird habitat on the upper San Pedro River (Udall Centre 1998). The CEC asked the Udall Centre to facilitate a public response period to the report written by the advisory panel of experts, and the panel developed policy recommendations, which included coordinated water resource management of the transboundary basin (San Pedro Advisory Panel 1998).

## Water Issues

Several issues of water allocation, demand, and quality create challenges and conflict.

### *Multiple Uses*

More recently, water-allocation issues surrounding human and environmental uses have become critical concerns and have sparked divisiveness among water users and water-management entities. Agriculture, cattle grazing, mining, and recreation remain the predominant land uses, though they are being supplanted by increasing urbanization. Additionally, as one of the most ecologically diverse areas in the Western Hemisphere, the basin contains as many as twenty different biotic communities and supports a number of plant and animal species of special concern to both countries.

The SPRNCA, an area of approximately 7,280 acres (18,200 hectares) managed by the U.S. Bureau of Land Management, is located north of the border. This strip is a major North American migratory bird corridor used by more than 350 species (CEC 1999; Liverman et al. 1997). The area has also been the object of an extensive series of observation and modelling activities by the Semi-Arid Hydrology and Riparian Areas Project (SAHRA) at the University of Arizona, the Semi-Arid Land-Surface-Atmosphere Program (SALSA), the Southwest Centre for Environmental Research and Policy (SCERP), and the Instituto del Medio Ambiente y el Desarrollo Sustentable del Estado de Sonora (IMADES), among others.

### Demand

For the basin as a whole, most of the water demand has been for mining, municipal and domestic use, and irrigated agriculture. Recent research suggests that riparian vegetation a/so requires a large portion of the water budget. In northern Mexico predicted decline in water availability due to climate variability may exacerbate competition for water resources between productive sectors, such as agriculture and industry, and domestic consumption (Magana and Conde 2001: 1). Currently the basin's water supply is considered to be in deficit, with annual withdrawals exceeding recharge by approximately six to twelve million cubic meters.

Increased production of copper from extensive ore reserves in Mexico limits groundwater availability for municipal and agricultural uses in that region and compromises water-conservation efforts. Expansion and modernization of the Cananea mine, particularly of the new concentrator, from 1978 to 1986 and again between 1992 and 1997, increased water extraction from 12.9 million cubic meters in 1980 to 20.2 million cubic meters in 1989 and 18 million cubic meters in 1990. On the U.S. side of the basin, total water extraction was 12.2 million cubic meters. Pumping in Sonora between Naco and Cananea in 1986 was estimated at 11.28 million cubic meters, but forty-eight wells were drilled between 1986 and 1994 to increase the water supply for mining operations, building up the pumping capacity to 40.2 million cubic meters (Arias 2001: 210). The Comision Nacional de Agua (Mexican National Water Commission) has recommended reducing the mine's use of fresh well water (SIUE 1993: 19, 76). Population projections for southeastern Arizona parallel those elsewhere in the Southwest—with roughly a 40 percent increase anticipated from 2000 to 2030 (Arizona Department of Economic Security 1997)—and will result in a major rise in water use to support subsequent municipal and domestic needs.

### Quality

In addition to the potential for water scarcity associated with overextraction and climate variability, groundwater and surface-water contamination also affects the quality of potable water supplies near the source of the San Pedro River. Inadequate (Naco) or nonexistent (Cananea) wastewater-treatment plants contribute to uncontrolled discharge of residual waters into the river. Unlined landfills introduce a variety of known and unknown substances that infiltrate into the aquifer. Moreover, the copper mines produce industrial waste that contaminates groundwater supplies via unlined and occasionally overflowing tailing dams. With the approval of the municipalities of Cananea and Naco, Sonora, and the support of the International Boundary and Water Commission (IBWC) and its Mexican counterpart, the Comision Internacional de Limites y Aguas (CILA), the University of Sonora's Department of Scientific Research and Technology (DICTUS), and the Arizona Department of Environmental Quality (ADEQ) conducted water-quality tests of the San Pedro River from 1997 to 1999.

Results indicated the presence of raw sewage and mining by-products, including cadmium, chromium, copper, iron, manganese, nickel, and lead, near the headwaters of the San Pedro and in wells close to Cananea. Further studies are needed to detect possible health problems due to the accumulation of heavy metals and sewage (nutrients) in people living in communities located along the San Pedro River. In the U.S. portion of the basin, numerous projects, SALSA and SAHRA particularly, have been conducting research on the hydrogeology of the area and the water needs of the SPRNCA. The purpose of this research is to provide information essential to the watershed organization, the Upper San Pedro Partnership (hereafter the Partnership), in constructing a water-conservation plan. The military base has lowered its water use considerably, and other conservation measures are being implemented. However, both the military base and the SPRNCA, as reserved areas, have prior rights claims to basin water, and the community in general and the Partnership in particular face both environment versus development and rural versus urban water conflicts.

These conflicts are complicated by binational suspicion. Mexican communities commonly think Sierra Vista wants Sonora to conserve basin water in order to promote more urban development, while some Sierra Vista residents believe that the Cananea mine's increased water use will dry up the SPRNCA and that its untreated sewage and heavy metals could contaminate San Pedro water flowing into the United States.

## Organizational Constraints

### *ARASA*

In 2001, a diverse group of stakeholders created ARASA, or the Sonora-Arizona Regional Environmental Association, in Sonora,

Mexico. The founders included teachers, doctors, mining engineers, attorneys, farmers, ranchers, and other citizens from Cananea and Naco, Sonora, as well as a small number of participants from Arizona, who sought to address regional environmental issues in Sonora. ARASA has defined its objective as being "to carry out actions that benefit the environment and at the same time improve the quality of life, through projects and actions oriented towards the protection, preservation, education, and scientific investigation of ecosystems and populations in the northeastern region of Sonora and southern region of Arizona." Additionally, members created an organizational philosophy of respect, tolerance, honesty, and professionalism.

ARASA's success has been impeded by a number of factors common to developing grassroots efforts. First, it has experienced difficulty narrowing its focus. Issues under consideration range from forest fires to environmental education and from biodiversity to industrial waste, to name just a few. Until recently, ARASA was unable to limit itself to a small number of issues it could begin work on and consequently establish its viability within the community. Similarly, insufficient organizational skills among the leaders has detracted from ARASA's effectiveness and deterred participation from citizens and state and regional agencies. Some members complain that, "They [ARASA] talk more than they act." However, given the fact that ARASA has been organized for only a few years, it has been remarkably successful in obtaining funding, constructing an organizational infrastructure with subcommittees, and selecting work activities tied to its objectives, including working with its U.S. counterpart, the Partnership.

At the same rime, poor access to scientific information has limited ARASA's capacity to carry out projects. A baseline of scientific evidence is essential. Without such information, ARASA's work cannot be fully legitimized in the eyes of local industries, federal and state agencies, and local communities. With this in mind, ARASA has recently established a technical subcommittee to collect and interpret current research on the hydrogeology of the basin and on land-use changes.

Finally, the search for funding has delayed project implementation, the development of group infrastructure, and the ability of the administrative council to move forward. ARASA has obtained grants from the Mascarenas Foundation and Foundation Mexico for Conservation, donations, and in-kind contributions by members. While ARASA now has office support, meeting space, and funding for environmental education, implementing projects will require additional money. Furthermore, ARASA depends on council members to volunteer their time, but many of them have one or two jobs along with ARASA responsibilities. Still, ARASA has defied heavy odds simply in coming into existence: no real government support, a powerful mining coalition, few resources, and a town fearful that environmental reform would threaten economic survival.

### *Upper San Pedro Partnership*

In 1998 the CEC initiated a technical study of the effects of water withdrawals on the international flyway along the river in response to a lawsuit filed by the Centre for Biological Diversity. "Over strenuous objections by some local officials, property-rights advocates, and anti-United Nations activists," a binational team completed the study and issued a report, Ribbon of Life: An Agenda for Preserving Transboundary Migratory Bird Habitat on the Upper San Pedro River (1999). The CEC, recognizing the importance of public response to this report, requested that the Udall Centre conduct a public-input process with basin stakeholders and residents. One of the report's recommendations was that multi-stakeholder watershed initiatives be created to help watershed decision making.

One year later a binational conference, "Divided Waters, Common Ground," was held in the basin, and a group of U.S. basin stakeholders began meeting in Sierra Vista. Within a year this group constituted itself into a watershed initiative, the Upper San Pedro Partnership. The Partnership's mission was "to coordinate and cooperate in the identification, prioritization and implementation of comprehensive policies and projects to assist in meeting water needs... to protect the people and natural resources of the Sierra Vista Sub-watershed... [and] to ensure an adequate long-term groundwater supply is available to meet the reasonable needs of both the area's residents and property owners (current and future) and the San Pedro Riparian National Conservation Area (SPRNCA)" (Upper San Pedro Partnership 2002: 2). The Partnership has been conducting research on basin geology, quantifying the hydrologic cycle in the basin, and developing hydrologic models in collaboration with other scientists. While the research is not yet completed, preliminary information has enabled the Partnership to begin developing a conservation plan to eliminate deficit water use in the sub-basin.

The Partnership Advisory Committee meets monthly in a venue open to the public and has formed its Staff Working Group whose chief concern is the conservation plan. The chairperson conducts meetings with a clearly established agenda. The Partnership has developed a $34 million five-year plan that pools financial resources from federal and state agencies.

Looking at both of these sub-basin groups, we can easily see that the Partnership has had three years longer to evolve and to acquire funding for projects. Similarly, the Partnership has established an organizational structure with subcommittees working on specific tasks, while ARASA has just begun to do this. Each group benefits from strong leadership from dedicated local people knowledgeable in policy issues and community concerns. Fortunately both ARASA and the Partnership have as part of their goals strengthening collaboration. The two organizations have exchanged letters indicating their desire to work together on a shared agenda regarding the exchange of scientific information and discussion of water-conservation strategies. The Dialogue on Water and Climate provides a forum for the groups to meet and discuss potential coordinated management projects.

## Representation of Stakeholder Interests

### *ARASA*

ARASA has developed a solid base among teachers, ranchers, and three Mexican environmental NGOs. However, membership and attendance suffered when ARASA became involved in a political controversy in 1999 over the attempted creation of a Mexican San Pedro Reserve that set rural landowners and the mining industry against regional environmentalists and SEMARNAT, the Mexican federal environmental agency. SEMARNAT has been managing the existing Ajos-Bavispe Reserve, which former director Julia Carabias Lillo wanted to extend to include portions of the San Pedro basin.

Carabias and former Department of the Interior director Bruce Babbitt signed a letter of intent to create this new Mexican reserve, called Mavavi. However, Sonoran mining companies feared their water supply would be cut off or limited, and ranchers feared their lands would be taken over and managed by SEMARNAT. Since ARASA's founders had advocated the creation of a San Pedro reserve, their role in the Mavavi flare-up placed them in a politically volatile position in the region. While this controversy has not been fully resolved, ARASA membership does include rural landowners and mining company representatives.

### The Partnership

Membership in the Partnership's Administrative Committee (PAC), which handles financial decisions, is open only to funding agencies. Membership on the PAC consists of agency heads, elected municipal leaders, two environmental NGOs, and a water company. While other stakeholders may attend and speak during PAC and Staff Working Group meetings, they do not have decision-making authority. When the group first began to meet publicly in 1998, there was considerable tension between city council members and Partnership members, but the Partnership has since admitted city council members and worked out many of the early differences over its responsibilities and power. The relationship between elected officials and other members of the Partnership may be tested once again when the Partnership completes its conservation plan and attempts to obtain public support for it.

Of the two San Pedro watershed groups, the Partnership is more heavily represented on the government agency side, while ARASA represents a strong grassroots effort, including three Sonoran environmental groups. At the same time, Mexico historically has lacked funding and governmental support for watershed councils (consejos de cuenca) along the border. It is that much more surprising, then, that SEMARNAT challenged Mexican San Pedro inhabitants to form their own environmental group (R. Barba 1999). ARASA took up this challenge and positioned itself as a grassroots organization with regional support. Now it holds public meetings with open doors and more private executive council meetings.

### Linkages

Currently ARASA seeks, though a technical subcommittee, a scientific assessment of current geohydrologic conditions in the entire basin, and is in the process of compiling research from both sides of the border on this topic. As part of this effort, ARASA is participating with the Partnership, the Climate Assessment of the Southwest (CLIMAS), and the Udall Centre for Studies in Public Policy in the Dialogue on Water and Climate (DWC), which interprets scientific research about the region and considers how this information might help water stakeholders, including managers, understand and establish an effective coordinated watershed management plan. Outreach to community members is one of ARASA's goals, and the group is designing an educational program for schools and the community.

ARASA's strongest card may be its capacity to initiate a community environmental education program. ARASA members have already had experience with educational programs in the local schools and are working with environmental NGO La Red Fronteriza del Medio Ambiente y Salud (Borderlands Network on the Environment and Health) which specializes in environmental education projects along the border. At the same time, while several of ARASA's members have science backgrounds and some policy experience, ARASA has only just begun to establish a baseline of scientific knowledge about the basin. Linking ARASA and the Partnership is essential, initially to share scientific research occurring in the Sierra Vista sub-basin, and later possibly to coordinate resource management. Interpretation of this research is essential for both community understanding and for water managers and city planners.

The Partnership has had the legislative support of Representative Jim Kolbe in attracting the funding necessary for geohydrologic research in Arizona. Its success in obtaining financial support was also due to the fact that it was one of the earliest of the Arizona Rural Watershed Initiatives to request funding for watershed research, so that the available funding pot was much larger. If the Partnership were starting now, it would have seventeen other watersheds to compete with for funding. Early on, Partnership agencies developed ties with the Southwest Watershed Research Centre and SALSA researchers at the University of Arizona. These researchers have helped the Partnership answer questions about water needs and existing and potential water supply in the Arizona basin.

In addition, while the Partnership's outreach program is just getting established, the group has had the advantage of an existing community environmental education program stressing water conservation (Project WET, also linked to the University, of Arizona's extension in Sierra Vista).

## Implications

The differences in these two watershed initiatives have several implications for coordinated binational watershed management:

- Historical knowledge of stakeholders' perspectives in this basin helps us understand how U.S. and Mexican stakeholders may be suspicious of each other's motives for participating in coordinated basin management. Mexicans claim upstream water rights, unlike the case in the Rio Grande or Colorado

River, and see water as key to regional economic development. Downstream Sierra Vista has historical reasons for being anxious about the quality and quantity of water flowing from Mexico. The lack of economic and institutional parlor between the two countries has also contributed to suspicion.

- Water issues in the two portions of the basin are different. Mexican residents have problems with water delivery and quality. Arizona inhabitants are concerned about a water deficit. However, these differences could provide bargaining strategies for both sides. Cananea and Naco need help with wastewater treatment, which could become recharge to the basin, thus potentially augmenting water available in the Arizona portion.
- Organizational constraints in the two parts of the basin could likewise be converted to strengths in coordinated management. The current DWC attempts to link the Partnership and ARASA in terms of research needs. Both organizations have geohydrologic information that the other would like to learn to improve local planning, especially with regard to decision support system models for water planners. Other opportunities exist for technology transfer, especially regarding water conservation, from Arizona to the United States.
- Watershed councils or groups should be truly representative of their stakeholders in order to obtain community acceptance and the political power to implement their decisions. Both the Partnership and ARASA could learn from each other about representation. We hope that the DWC can be a means of this occurring.
- Collaboration among scientists, policymakers, and stakeholders varies from country to country, but in the case of the San Pedro, ARASA is very interested in strengthening its ties to university researchers and stakeholders within Sonora. ARASA views the Partnership's success as being linked to its university connections. However, ARASA is suspicious of Mexican government participation in their efforts. The group reels its efforts may be co-opted by state or national agencies, as has often happened in the past.
- National policy considerations influence the potential for coordinated basin management. Local initiatives along the

northern Mexican border are linked to national policy demands. Mexican environmental policy frequently runs counter to Mexican economic policy in the critical importance attached to development, especially in mineral resources and maquiladoras along the northern border. One study of textile enterprises indicated profits were more important than environmental protection to 46 percent of the managers. Second, SEMARNAT receives very limited resources that must be allocated among too many programs for it to effectively manage environmental policy (Romero Lankao 2001: 176-178). At the same time, within SEMARNAT the Comision Nacional de Agua has instituted a new "culture of water" and provided guidelines for restructuring the management of aquifers through watershed councils. These watershed councils are intended to link state and municipal government with local community participation in managing and financing systems for potable water, sanitation, and irrigation.

- Because the northern frontier is characterized by rapid demographic and industrial development, the border has its own water program emphasizing conservation of ecosystems, reversal of industrial contamination, public participation, and environmental education. This program promotes binational environmental educational programs and information exchange. While the role of environmental NGOs in the consejos is not discussed directly, SEMARNAT policy programs all clearly advocate the participation of civil society, including environmental groups. However, no indications of government funding support exist, except for one education program in 2002 and six planned by 2006.

What is so significant about the potential of the binational resource alliance? While the San Pedro River is not on the same scale as the Euphrates, the Colorado, or the Rio Grande, this alliance between its two watershed organizations provides the opportunity to link science with water management and policymakers. While basin research has been extensive in the U.S. portion of the basin, scientists on both sides realize their findings need to be integrated in order to achieve a more accurate and useful picture of watershed supply and demand. Without this integration, basin water managers and policymakers are handicapped in their planning efforts and opportunities for effective conservation strategies are more limited. The San Pedro DWC provides opportunities of scale in that hydrologic models and decision-support tools are being constructed for binational use. Transboundary, rather than separate, management strategies are especially important in a situation where what Mexican stakeholders do has the potential to affect what happens in the U.S. portion of the basin.

High stakeholder involvement, particularly on the Mexican side, increases the potential for success in any watershed initiative. Public debates within these two watershed organizations have "stirred controversy and revealed the importance of accounting for the region's social and political forces", but this has been a necessary step in the process of educating water managers, municipal officeholders and agency representatives, as well as basin residents in general, about the science of basin water management. While this process of public debate and planning, assisted by the recently initiated DWC, will continue for years and probably decades, current Mexican and U.S. water policies encourage the linkage between grassroots stakeholder organizations, governmental agencies, and scientific research as a more effective management tool than the old top-down model. The Partnership has been receiving ample funding and legislative support for basin research, and ARASA has been successful in obtaining capacity-building and program grants within a relatively short period of time.

Finally, the diversity of administrative issues and cultural concerns in tiffs case provides a good model of what goes on along the U.S.-Mexico border. The upper San Pedro River basin offers an opportunity of scale in that complex water issues are being addressed by scientific research and policy debate within a compressed area. As a result, the upper San Pedro has become a demonstration basin for the North American HELP and DWC projects with the idea that water managers and policymakers, especially the majority confronted with transboundary water management, can learn from each other's experiences.

## Western Water Policy Reform

Water policy has left an indelible mark on America's western landscape. During the settlement and development of the West, a set of water policies evolved to create institutions and enable construction of facilities that allocate, store, distribute, and manage water. The most stunning tangible manifestation of those policies is the built environment. Though less visible, the underlying legal and institutional framework also stands firm with its foundations deeply set. Either the built environment or the laws and institutions would be extremely difficult to remove or fundamentally change, but each must be reformed and operated differently to respond to the changing nature of the West. There is no better example of the impacts of water policy, and no clearer illustration of the needs and possibilities for reform, than the Columbia River.

The Columbia is one of the world's great rivers. It drains an area the size of France and has more than twice the flow of the Nile. Explorers Meriwether Lewis and William Clark wrote in wonderment about the hordes of fish upon their first view of the upper reaches of the Columbia. In 1805, they traveled down the Columbia to the ocean without a single obstruction. The team only had to portage around Celilo Falls, the great narrows where the region's tribes speared migrating fish and sustained a satisfying life and rich culture.

## A Monument to Western Water Policy

Today, the Columbia River has the distinction of being the most developed river in the world. Only one fifty-mile stretch of its twelve hundred miles remains "undeveloped"—less than five percent of its length from the Bonneville Dam to the Canadian border. Long steps of still water lie end to end. There are seventy-five major dams in the Columbia River system, including fourteen constructed in the mainstem of that mighty river. The great hydro-electric dams of the Columbia enslaved the river for the sake of aluminum plants and hair dryers; millions of people now depend on the system for their electric power. Each year, irrigators divert from the Columbia an amount of water greater than twice the entire annual flow of the Colorado River turning deserts upstream at the base of the Rockies into gardens. The river also has been pressed into service to carry away wastes—sewage, toxics, salt, and silt.

As throughout the West, the laws and water projects on the Columbia have spurred progress. For example, water project construction created jobs, and low power rates enhanced business expansion. Also, greater crop yields through better irrigation enriched communities. A Bonneville Power Administration publication a few years ago bragged that [i]n little more than one generation Man has harnessed the tremendous water power of the Columbia Basin.... e has tamed floods, improved navigation, and turned deserts into rich farmland.... [P]roduction of low-cost electricity has been a major factor in the Pacific Northwest transition from a regional economy based on agriculture and lumber to a more balanced, widely diversified economic and social structure.

This progress created costs and impacts that were largely ignored at the time the water laws were written and projects planned. The inexpensive electric power produced by the Columbia River system proved to be not so cheap after all. Although the price was low, the costs were high. Power only seemed cheap because consumers did not have to consider the inherent value of the salmon and the tribal societies that depended upon them. Consumers also could ignore the value of ecosystems, free-flowing rivers, and lost gene pools.

In recent years, billions of dollars have been spent and committed to rescuing the Pacific salmon. Desperate efforts have included barging and trucking migrating juvenile fish around the dams, building fish ladders and elevators, and replacing waning natural fish populations with hatcherybred substitutes. Meanwhile, plans are being drawn to sacrifice water, stored above dams for lucrative power generation and irrigation, by releasing it to restore some semblance of the river's natural flows during the times of the year when fish would benefit most.

The most notorious tragedy of development on the Columbia is that three-quarters of the historical salmon population no longer can survive in the river. Fish used to travel as many as nine hundred miles up the Columbia and Snake River system to reach their spawning grounds. While much of their spawning habitat far upstream is in good condition, with some of it in wilderness areas or wild and scenic rivers, salmon can no longer get to these waters. Today, one or two surviving salmon, the last vestiges of their race, make news when they succeed in their struggle back to Redfish Lake in Idaho.

As the Ninth Circuit said last year, "it is generally accepted that the Basin's hydropower system is 'a major factor in the decline of some salmon and steelhead runs to a point of near extinction.'" However, big dams are not the sole reason that the salmon cannot survive. Fish habitat and migration are also threatened by small dams, stock-watering ponds, miners' tailing ponds, and irrigation diversion structures that glutted headwater streams and flooded spawning beds.

With the demise of salmon populations came the demise of a lucrative and active commercial fishing industry that depended on the

Columbia River fisheries. The loss of fish-related jobs and businesses, often in families for generations, forced dislocations and suffering upon many communities. Perhaps most striking was the impact on traditional tribal societies. The salmon are the "buffalo" of the Northwest Indians. Tribal society was tied to subsistence, commercial, and spiritual relationships with salmon, similar to the relationships that existed between the buffalo and the Plains Indians. However, there are differences. The buffalo had virtually disappeared by 1883; it has taken an additional century for salmon to be brought to the brink of extinction. Destruction of the buffalo was a purposeful enterprise for non-Indian society; destruction of the salmon was less purposeful, but it seems destined to become as severe as the near-extinction of the buffalo. The consequences for the tribes in both cases go well beyond the economic effects, eroding the very base of culture and community.

The system of hydropower dams that was heralded as a great economic boon to the region is viewed these days with colder, more discerning eyes. Similarly, the big irrigation diversions and uses of the river, like logging and manufacturing, brought benefits to the area, but they can no longer be considered apart from their negative effects on the river's health. Clearly, the elaborate hydropower system on the Columbia was vastly over-built. It will never again be operated at its full, power-generating capacity because of the incredible destructive potential that the system holds. This failure has resulted in the current situation in which few people are seriously considering the eleven new hydro dams that have been proposed for the Snake River System.

Water law and water policy are ultimately to blame for permitting and even encouraging what has happened to the Columbia system. Yet, I believe that the trend can be reversed on the Columbia and other western rivers by redirecting the traditional instruments of water law and policy—beneficial use, water projects, and watershed management—to serve modern values and fit modern conditions.

## Fundamental Principles of Water Law

Many western state statutes declare that "[b]eneficial use shall be the basis, the measure and the limit" of private rights in water. Among people who compete to put water to a beneficial use, the earliest users have the best rights. These simple rules sum up western water law, but their application has varied with changing conditions in the West. What constitutes a beneficial use necessarily evolves over time with the needs and values of society. In the early days, the West was undeveloped, and water was relatively copious, though not evenly distributed. Water could be distributed in the best interests of society by allowing it to be committed to categories of use, like mining or agriculture, that were generally considered beneficial. Very early, it became necessary for the courts to compare uses and means of diversion to determine "reasonableness" and to weigh their relative efficiencies in accomplishing beneficial purposes.

The first cases that arose dealt with the prevention of "waste." As one early case said, [i]t is elementary that the waters of the public streams of this state belong to the people, and that appropriators acquire only a right of use. It is also settled law that an appropriator is limited in his use of water to his actual needs. He must not waste it.... A recent case has held that "[t]he owner of a water right has no right as against a junior appropriator to waste water, i. e., to divert more than can be used beneficially." Therefore, beneficial use is a criterion that limits the amounts and uses of private water rights.

In 1993, the Washington Supreme Court was confronted with an appeal by Clarence and Peggy Grimes, who contested a referee's ruling in an adjudication that curtailed their right under Washington law to divert 3 cubic feet per second (cfs) based on uses going back to 1906 to only 1.5 cfs and limited their 1520 acre-feet storage right to 920 acre-feet. The court upheld the reductions, stating that "[t]he key to determining the extent of plaintiffs' vested water rights is the concept of 'beneficial use'.... An appropriated water right is established and maintained by the purposeful application of a given quantity of water to a beneficial use upon the land." However, the "amount of water necessary for a beneficial use" will be limited to a "reasonable" amount for the particular purpose.

The early cases dealing with beneficial use (and the reciprocal concept of "waste") were decided in a simpler era than ours. Water was more plentiful and no one spoke up for fish or natural systems, let alone the raw beauty of a place. Today, the number and variety of competing uses have proliferated. Water users and water rights holders include fishers, boaters, and environmentalists. Although they are not always heard, there are also groups, like WaterWatch of Oregon, who regularly raise issues concerning water quality and ecological integrity. Modern courts recognize the force of changing values and expanding uses. In 1992, the Wyoming Supreme Court said that "'[b]eneficial use' is... an evolving concept and can be expanded to reflect changes in society's recognition of the value of new uses to our resources." Similarly, the Idaho Supreme Court has said that "[w]hat is a beneficial use, of course, depends upon the facts and circumstances of each case." And in Washington Department of Ecology v. Grimes, the Washington Supreme Court linked the idea of beneficial use with other uses that may be possible: "A particular use must not only be of benefit to the appropriator, but it must also be a reasonable and economical use of the water in view of other present and future demands upon the source of supply."

State constitutions and statutes in the West almost uniformly dedicate water to the public. Accordingly, private water rights can be granted when the use is consistent with the "public interest" or "public welfare." The public interest provisions in state constitutions and statutes potentially allow for a comparison of various benefits, leading to limitations on those that are less socially beneficial. For example, decision makers might opt for limiting agricultural uses to promote municipal uses, or fish and wildlife purposes might take priority over industrial or power uses. If water is a public resource, public agencies should have the power to decide whether a use is beneficial by balancing the consequences of a proposed private use of the water with a broader public interest. Officials must construe the meaning of "beneficial" in its full state law context. The context is one in which a right to water is being transferred from the public to private hands. The public trust doctrine expresses essentially the same notion: The public has a stake in all unappropriated water that cannot be defeated by official neglect.

When applied to a request for new water rights, a request for changes of uses, or any other opportunity for state review, the doctrine of beneficial use should be an engine of the public interest. A progressive concept of beneficial use should influence all water decisions, including the hundreds of applications for new water rights pending in Oregon, Washington, and Idaho.

### *Turning the Curse into a Blessing*

After the turn of the century, states gladly let the federal government introduce national programs and supplant local initiative and traditional state control of water. States were seduced by the idea of big spending and dreams of burgeoning economies. However, the price of receiving these subsidies included loss of state and local control as well as dramatic physical changes in the local environment;

the Bureau of Reclamation (BOR) and U.S. Army Corps of Engineers (Corps) took over responsibility for planning and allocating much of the West's water. In its ninety years, BOR alone has constructed over six hundred dams and sixteen thousand miles of canals, attempting to correct the errors of nature's ways by delivering water where and when it was demanded. Nowhere is the handiwork of federal water development more impressive than on the once mighty Columbia, the nation's second largest river.

The Columbia River's plumbing system is a spectacular feature of the Northwest's built environment. Unfortunately, this wonder of engineering turned out to be devastating for the natural environment, especially the salmon fisheries. When the National Marine Fisheries Service (NMFS) issued a biological opinion in 1993 stating that the Columbia River power generating system could operate without jeopardizing the Snake River salmon in violation of the Endangered Species Act (ESA), the State of Idaho challenged the ruling. This challenge resulted in the federal court throwing out the NMFS opinion, because it found that the federal agency had used data and modelling methods selectively in order to minimize the likelihood of salmon extinction.

It is common in the West to think of Corps and BOR facilities as causes of environmental problems, not solutions. Dams and diversions are clearly "Public Enemy Number One" when it comes to destroying the salmon runs of the Columbia Basin. Throughout the West, big mainstem dams are arguably monstrous destroyers of habitat and recreational opportunities. Anyone who has floated the Colorado River knows that there may or may not be a beach to camp on for the night because of power operations in the huge Glen Canyon Dam. The dam has stopped the flow of sediment through Grand Canyon and caused the water to be sent through the canyon in sporadic cycles corresponding to the demands of electricity users. The river may be low and slow when you roll out your sleeping bag at dusk on a vestigial sandbar. By dawn, a raging river may be lapping around you.

A more optimistic way of thinking about the built environment, including the complex of dams, canals, and irrigation systems, is to consider these facilities as resources to be manipulated to satisfy current societal values. The Natural Resources Law Centre (NRLC) at the University of Colorado School of Law has recently completed a study for BOR and the Environmental Protection Agency showing the potential for managing BOR facilities for ecosystem benefits.

Managers of several reclamation projects around the West are actively exploring new methods of operation and achieving new, publicly beneficial purposes never anticipated during the authorization and construction of these projects. In the upper Colorado River, loss of indigenous endangered fish species habitat is a serious problem that is being addressed, in part, by releasing water from Flaming Gorge Dam on the Green River in Utah and Blue Mesa Dam on the Gunnison River in Colorado. The quantities and times of releases correspond with the spawning and habitat needs of endangered squawfish, humpback chub, and bonytail chub.

The endangered species problem in the Colorado River is also addressed as a by-product of salinity control programs that were designed to correct another environmental phenomenon exacerbated by the dams. Irrigation return flows were turning the river downstream so saline that water was becoming unusable for irrigation. BOR launched multi-million dollar projects to reduce salinity levels and attack causes of salt loading. Lining irrigation ditches and making the systems more efficient has reduced the salt loading as well as the quantity of water that needs to be removed from the river. As it turns out, keeping water in the river also benefits the fishery.

The upper Snake River provides another example. The Snake in Idaho was historically the richest habitat for salmon in the Columbia system, but is now virtually barren. At least seven major reservoirs are strung out along the upper Snake. If they are operated so that the releases downriver are in sequence with the needs of salmon, they can be used to mitigate and in some cases enhance salmon migration. The solution is complicated because BOR cannot simply open and shut at will the outlet works of dams that it owns. There are contracts for the delivery of water for irrigation and hydro-electric power generation. Thus, BOR has to pay for water it uses to maintain streamflows for fish.

There are other problems in using facilities like those on the Snake to restore conditions needed for salmon. The amount of water needed for the river and its fishery may be solved by releasing water, but salmon respond to a particular water temperature that provides signals to them about when and where to go. Water stored at the bottom of a reservoir is colder than the water at the top. Some dams in the reclamation system were built to be drained fully, with outlet works at the bottom, so that cold water is typically released. Others are built for hydro-electric generation with high outlet works that provide warmer water. These designs may or may not coincide with temperature needs for fish. Modifying the dams is an expensive proposition. For instance, to change the release system at Shasta Dam on the Sacramento River in California for salmon needs will cost $80 million.

Progress in using BOR facilities to accomplish broader ecosystem benefits is shown by the Yakima Project on tributaries to the Columbia in Washington. The Project enjoyed impressive crop yields with a four hundred percent increase in irrigated acreage. Like so many of the Yakima Project's relatives throughout the West, however, single-minded efforts to improve agricultural yields limited both the vision and reality of the project.

In the Yakima River Basin, salmon populations have declined by approximately ninety-nine percent since the turn of the century. The first wave of efforts to protect salmon involved the construction of fish ladders and passageways around the dams and screens across diversion structures. These efforts were limited in their success and could never counter the assaults on natural systems from dams that allowed over-appropriation of water and placed multiple obstacles in the way of migrating salmon.

The pressure to improve river conditions to meet the needs of salmon was given legal force by the recognition and enforcement of Indian treaty fishing rights. Several years ago, the Yakama Indian Nation received an award of over $2 million from the United States for the destruction of the tribe's treaty-secured fishing rights by the Yakima Project. Recently, the state court ruled that the Yakama Nation has certain minimum streamflow rights that are necessary to sustain the treaty fishery. Therefore, the Yakima Reclamation Project must be operated to ensure the instream flows necessary for the tribe's fishery.

Litigation and lessons learned from destroyed habitats and fish runs in streams periodically dried out by irrigation demands expanded local consciousness. People in the Yakima Basin began to search for ways to ensure that' secure water uses could continue while fish habitat was improved. Now, BOR and water users have instituted a "flip-flop" method of operation. Under BOR's former operations, the Cle Elum Dam held back water when irrigators did not need it and released it in huge amounts when irrigators called for it. This resulted in keeping water away from salmon spawning beds in the winter when the nests of salmon eggs (reads) needed to be washed over with water.

Then the stream had artificially high flows during the irrigation season, sometimes destroying the reads. Now, by operating this dam in conjunction with another dam on a different arm of the Yakima System, releases can more closely mimic nature, and water deliveries can be satisfied by releasing water stored on the other tributary.

Water resources in the Yakima Basin remain over-committed, following a century of later commitments that conflict with treaty obligations to the tribes. However, recent efforts are allowing tribal and irrigation rights to be satisfied. Currently, the dams can be operated to ensure that tribal rights to water and fish are fulfilled. However, this requires the acquisition of supplemental water. To address that need, a watershed council in the Yakima Basin is now considering proposals either to create a water banking scheme or implement a water leasing and transfer program.

Furthermore, recent congressional legislation enables more flexible use of BOR facilities on the Yakima, holding out hope for the fish and tribal fisheries. For instance, the Yakima River Basin Water Enhancement Project Act authorizes federal funding to purchase or lease water directly, which would make the proposed water leasing or banking programs possible. The project is not perfect. The price tag is high, and the solutions challenge engineers. However, the required investments are necessary to complete the Yakima Project by bringing it to the point that its operations can be more or less sustainable.

## Watershed Governance: From Political to Natural Boundaries

Agencies, districts, and cities that develop and supply water typically have jurisdictional boundaries that bear little relation to the scope of the impacts of the water decisions made by these organizations. Also, the allocation of agency authority may be inappropriate. For example, water quality and quantity are usually regulated under different laws and separate agencies. Washington has been an exceptional bastion of sound judgment in this respect, but is considering a regressive bill that would put water allocation in an agency other than the Washington Department of Ecology, where it is now wisely lodged along with water quality regulation.

Groundwater and hydrologically connected surface water are often subject to different laws. Although some states are considering proposals to reverse this ignorance of hydrology in favour of conjunctive management of surface and groundwater, others are moving in the opposite direction. For instance, the Oregon Legislature was actually considering a bill to segregate the Oregon Water Resource Commission's consideration of surface water from hydrologically connected groundwater.

People with interests affected by water decisions are frustrated with the water decision-making processes in state agencies that exclude their participation or influence. State laws, such as those in Colorado, do not even allow consideration of factors related to the public interest when major water decisions are made.

One response to the public's frustration with water institutions has been to create parallel and sometimes conflicting institutions. More often than not, they are the result of federal legislation. Since the 1970s, Congress has passed an impressive body of laws that regulate the impacts of water use and development. They include pollution laws like the Clean Water Act and implementing statutes in each state. Although a few states have their own programs, wetlands protection comes mainly through a federal permitting program for dredge and fill operations. But these programs remain piecemeal, covering parts of water-related environmental problems in a rather uncoordinated way.

Federal environmental laws often import national standards inappropriate to the local situation. These laws sometimes sweep with too broad a brush, overlooking the peculiar needs of ecosystems and communities. They are also typically ineffective in dealing with cross-jurisdictional problems. Federal environmental laws are classic command-and-control regulatory programs that can operate in a heavy-handed way and are often, like litigation, confrontational and polarizing. There is a resulting backlash against regulation that has provoked efforts to overhaul environmental law. While there are legitimate criticisms, I do not believe that these laws are broken and in need of fixing.

Whatever their flaws, the federal environmental laws, not state legislation, have taken the lead in responding to public fervor for environmental protection. Insistent on maintaining autonomy in water matters, states have attempted to mute these federal laws by persuading Congress to insert language in them providing that nothing shall interfere with the states' rights to allocate water. Although the states could use their autonomy to tailor environmental controls to unique situations and water allocation schemes, they have been slow to address the kinds of public concerns that led to the enactment of environmental laws, especially in the field of water resources. Finally, states are beginning to rise to the challenge and recognize that water law must serve more than those who happened to get to the water first. A few years ago, a group announced the "Park City Principles,n which called upon states to "fashion water laws and institutions responsive to the entire range of water values and interests, including those not traditionally recognized in water law and administration." This group was not comprised of environmentalists or activists; it was a group convened by the Western Governors Association and the Western States Water Council, organizations that can properly be called the heart of the western water establishment. These organizations realized that "public values [are] now protected primarily under federal laws." Therefore, the Park City Principles implicitly accepted that states had effectively abdicated responsibility, and if states truly want to play the primary role in water management, they must assume commensurate leadership, authority, and accountability.

Some of the most promising responses to institutional problems in water policy have begun to come from the grass-roots. Locally-generated efforts are supplanting inadequate state allocation and regulatory systems. These efforts are bringing rationality to the application of federal laws. Consider the experience of people on the Henry's Fork of the Snake River. A few years ago, the tensions in this Wild West region of Idaho (and a little piece of Wyoming) were at a breaking point. Ranchers and farmers had long lived there, but with people coming from the cities to seek beauty and solitude, a tourism business was growing up and some people were building second homes. Grazing practices and irrigation return flows were at odds with fish habitat. The newcomers also favoured planning, while a local irrigation district preferred a laissez-faire approach. These two groups were on opposite sides of a proposed mandatory watershed planning bill before the state legislature. The bill was defeated, but no sooner were the warring neighbours back home when disaster hit. A construction accident at the Marysville Hydro Project resulted in the release of a huge amount of sediment into Fall River. At the same time on another part of the river, drought had lowered Island Park Reservoir, and managers decided it was good time to drain it, kill the "trash fish," and let some sediment out. The two surges of sediment-some 70,000 tons-choked the river, impacting water users and fish life. Everyone started blaming everyone else, but they soon realized that the problem may not have occurred and solutions would not have been nearly as difficult if they had been communicating with one another. Public meetings finally brought people together in crisis and anger. However, they talked and ultimately agreed to start the Henry's Fork Watershed Council. The two arch-rivals in the area, the Fremont-Madison Irrigation District and the Henry's Fork Foundation, a local environmental group, miraculously agreed to co-facilitate the new Council. In the next legislative session, the two groups who had been fighting with one another went back to the legislature, this time side by side. They asked for and received legislation that unanimously approved a. charter giving the Council a role in reviewing all agency proposals for any development or project in the watershed.

Cooperative ventures are springing up in watersheds all over the West. Exasperated that western water law, environmental regulatory laws, and state and federal agencies do not provide for their interests to be heard, respected, and reflected in the decisions, people are venturing to solve their own problems. NRLC has been studying these efforts around the West during the last two years and has collected the stories of groups in eighty different watersheds who organized and took charge of their own destinies,

Rather than waiting for someone from the federal or state government to come in and solve problems, local citizens are taking responsibility. In Colorado's Yampa River Basin, people are tackling issues that arise when entrenched farming and ranching interests conflict with interests in the growing area around Steamboat Springs. Concerns include recreation, established agricultural water supply rights, water quality, sewage disposal, an endangered fish recovery program, traffic, land use, housing, school crowding, and escalating social services costs. The formation of the Yampa River Basin Partnership grew out of a meeting that approximately 260 people attended in late 1994.

Some states provide frameworks and incentives for watershed-based efforts. In 1993, Oregon passed legislation encouraging the creation of watershed councils. Oregon is leading the way for other states by supporting local efforts to solve problems on the ground with its Watershed Management Program. One example of the Oregon program in action involves the Illinois River, a tributary to the Rogue. Initially, efforts to work together in dealing with problems of water quality and deteriorating fish habitat were hampered by the polarization of commodity and environmental interests. After the Oregon Legislature passed the Watershed Management Program, the local Natural Resources District convened people and applied for a state grant from a fund provided by lottery proceeds. In addition to district board members, the Illinois Valley Watershed Council now includes representatives of the fishing industry, educators, miners, environmentalists, and the City of Cave Junction.

Watershed-based decision-making tends to reflect greater diversity of public interests and can fine-tune solutions specifically to the affected place and people. The watershed is a flexible concept, because it is an amalgam of countless sub-watersheds nested together. Watersheds can be grouped or separated to define a geographic area that coincides with the sources and effects of a particular problem. Some have called this a "problemshed" approach. Therefore, if an area has a soil erosion problem, the problemshed might include land draining into a tributary where logging is taking place and an adjacent tributary where irrigators are returning silty return flow. The problemshed should include the area affected as well as the source of the problem. Thus, the drainage downstream from an erosion problem might be included if a town there draws its drinking water from the stream or a community that stakes some of its livelihood on the benefits of fishing from the stream. Larger or smaller problems can be addressed by adjusting the scope of the watershed.

The watershed ideal has succeeded where traditional water law and institutions have faltered. Can it work for big problems? What about a problemshed as big as the entire Columbia? Congress created the Northwest Power Planning Council (NPPC) in 1980 after government agencies and power interests had made disastrous, single-minded decisions that resulted in a financial crisis for the power generating complex on the Columbia. The power program was confronting bond failures for massive investments made in oversized, poorly designed nuclear power facilities and an ecological crisis created by a hydro-electric system that was being operated with insensitivity to the invaluable salmon stocks of the sprawling Columbia system. The world's greatest salmon river was crashing.

In an exercise of uncommon wisdom, Congress mandated that future energy planning for the Columbia system must be a process that includes the public and considers the full economic and ecological effects of various alternative sources of energy supply. Congress required the creation of a program to reverse the effects of the string of giant dams on the waning fishery. The Northwest Power Act has been somewhat successful, but the tragic condition of Northwest salmon fisheries overshadows the Act's success. Two hundred fourteen stocks are threatened with extinction, and fishing in recent years is close to a standstill in response to requirements of the ESA.

The old decision-making process regarding the Columbia had its faults. It excluded too many people and groups, and included a range of interests and values that was too narrow. This was typical of water planning in the past. The Northwest Power Act approach forced a wider, more integrated consideration of affected interests. It fell short, though, by failing to deal with a fuller panoply of influences on the life-cycle of salmon-issues like land use, water use, and water allocation. Also, the Ninth Circuit rejected the Council's Strategy for Salmon and chastised it for failing to establish leadership, thereby "sacrificing the Act's fish and wildlife goals." The Council has responded with a new plan that seems to be superior to the revised NMFS biological opinion that replaced the one the court found deficient. Angus Duncan, former chair of the NPPC, has proposed a watershed council for the Columbia that could be created by broadening the Northwest Power Act to consider land and water management, expanding the power and fisheries purposes of the present Act. It is a bold proposal that surely will be seen by some as threatening the institutional turf of existing agencies and governments, but it certainly deserves attention and serious consideration.

It is tempting to dismiss western water law because it has failed so many interests in the West, but that is neither necessary nor politically feasible. The beneficial use doctrine has been flexible in the past, and it can be in the future. Through that doctrine, the law of prior appropriation can assimilate today's vision of what is "beneficial." The physical reality of the vast plumbing systems that mark the western landscape is too great to ignore. Some dams can and must go, but the dams that will be removed are few and far between. The rest, including nearly all the big ones, cannot be wished away, and there is no monkey wrench big enough to fulfil Edward Abbey's fantasy even if one could discount the social and natural destruction that would go with removing all the big dams. They can be retrofitted and re-operated. That is feasible, and it is happening.

The idea of using watersheds as geographic units for solving natural resources problems is timely. They capture the appealing and currently popular idea of decentralizing political control. The beauty of these locally based solutions is that they call on people to exercise the responsibilities of citizenship-to participate in finding solutions and making them work. Fulfilling this fundamental idea of citizenship and civic action calls for unparalleled participation by members of the public in their professional and citizen roles. In this way, the public can take back the water projects, water policy, and the rivers. The public can change the course of the Columbia, for the people and for the fish. Water planners in the United States, seeking a new paradigm, could learn some lessons from their friends Down Under. United States water policy is in flux as it moves from the chief paradigm of the 20th century—multiple-purpose development—to one that seeks to use water in more environmentally sustainable ways within the constraints of existing allocations. As a result, the institutions that have managed and allocated this country's water resources are becoming strained and less able to perform their historic function of mediating competing demands for water. Water institutions, including legal institutions, must continue to perform many of their traditional functions and at the same time adapt to new demands such as environmental protection, ecosystem restoration, and social equity. Throughout the country, the great era of dam building is over, although the pork mill for navigation-improvement projects grinds on. Indeed, the end of the dam-building era heightens rather than lessens competing demands for water. Today, growing cities compete proponents of aquatic ecosystem restoration, who compete with traditional users such as agricultural irrigators. All parties compete among themselves for stressed supplies.

In the rapidly growing arid and semi-arid western United States, we have moved from the reclamation era—characterized by large, federally subsidized, regional water projects—to the era of reallocation, conservation, and aquatic ecosystem restoration. The new paradigm seeks to support the traditional consumptive uses such as irrigation, and nonconsumptive uses such as recreation, fisheries maintenance, and ecosystem restoration, through more efficient use and better management of demand. It also seeks to promote reallocation so that existing unregulated rivers and their ecosystems can be preserved, and degraded ones can be restored. The humid East also faces increased risks of water shortages in some rapidly growing areas, and it must deal with many of the same problems of restoring aquatic ecosystems that the West faces.

### *Stormy Weather*

Shifts in policy have sparked a contentious debate about how the risks of shortages should be shared among competing users. Moreover, disagreements abound over how the benefits of alternative schemes should be distributed. These problems have taken on an added complexity in the debate over global climate change. The hydrological, economic, and political consequences of global climate change in a given watershed or river basin are uncertain. Some predict that global climate change may alter precipitation and runoff patterns throughout the world. The rub is that both the amount and timing of rainfall may change but the geographic and temporal scale of the change is uncertain. Some regions, such as sub-Saharan Africa, may experience decreased precipitation and more-extended droughts. Others will see increased precipitation and more-frequent and more-severe floods.

Increased precipitation, however, may not translate into more-available water supplies in all regions. In water-short areas with historically variable rainfall patterns, increased precipitation may actually exacerbate efforts to provide reliable water supplies. Warmer average temperatures may cause spring runoffs to come earlier and evaporate faster, snowpacks may melt earlier, and more precipitation may fall as winter rain rather than snow.

Increased, but out-of-cycle, rainfall is the projected pattern for parts of the western United States. Wetter, warmer weather could strain existing storage systems that currently provide reliable regional water supplies. Existing reservoirs may not be able to capture the increased winter runoff, causing serious shortages in the summer. In addition, states and regions may have to adapt to a series of ecosystem changes due to plant and animal population shifts caused by changing climatic patterns. All these new uncertainties' must be factored into any adaptation strategy.

### *Policy Adrift*

New demand and greater uncertainty about available supplies mean today's water users and managers face difficult choices. These difficulties are exacerbated by increasingly thin institutional buffers between groups with diametrically opposed points of view. No single group, such as irrigated agriculture or municipal water supply, controls the agenda and the water resource agencies. The diffusion of power among old and new stakeholders means that agencies have increasingly less power to resolve conflicts by imposing a solution.

In addition, the two major institutional frameworks—state water law and federal water resources development—are under great stress. State water law has traditionally determined the allocation of water among competing residential, industrial, and agricultural users and hydroelectric power generators. But increasingly the market, rather than state water administrators, controls the allocation of water.

While the federal role remains important, it is changing and diminishing. John Volkman, a longtime student of the Columbia River—which has been primarily dedicated to hydroelectric power generation, navigation, and irrigation to the detriment of ecosystem services—has characterized the conflicts among interests in that basin as a contest between a working river and a river that works. Change is difficult on developed rivers because most users, especially established consumptive and nonconsumptive ones, view the allocation of water among competing users as a zero sum game.

In the past, when water policy meant developing national water resources, the federal government was able to solve most conflicts by simply providing more usable water by funding a large project. But modern water policy is no longer an important national political issue, and the federal government is now more of a regulator than a supplier. The new water era is therefore much more diffuse and decentralized.

At the federal level, the two major historic water agencies—the Bureau of Reclamation and the U.S. Army Corps of Engineers—face shrinking budgets. They must also share their authority with such agencies as the U.S. Fish and Wildlife Service, the National Marine Fisheries Service, and the U.S. Environmental Protection Agency, which are not water managers per se but do have considerable authority to influence specific management decisions—especially decisions that could alter habitats and aquatic ecosystems—and are becoming major players in water policy. While the Corps and the Bureau of Reclamation still have a political constituency, they lack much of the autocratic power that they once had. Much of the old planning infrastructure has been dismantled, and the Corps and the Bureau are caught between their old constituencies and efforts to expand their mission to include environmental management and conservation. The Bureau of Reclamation's draft strategic plan for 2000 through 2005, for instance, grandly proclaims that the Bureau's mission includes "managing, developing and protecting water and related resources to meet the needs of current and future generations.

### Hands off

The institutional context of the new era differs markedly from past regimes. For a variety of reasons, it lacks the clear, coherent vision and focus on multiple use. First, the new policy is still largely a negative reaction to the social and environmental costs of multiple-purpose development—especially large dams—and it will take time for a clear vision to evolve. Powerful new ideas are emerging, such as the goal of the normative river," which seeks to restore dammed and channeled rivers to a state closer to the original. These ideas are complemented by new management theories that consider ecological processes as continually evolving rather than reaching a permanent, natural equilibrium. These new approaches have great relevance for water management. But ecologists rather than engineers are the intellectual force behind the new vision, and their influence is still marginal. Second, in contrast to multiple-purpose development, the new policy lacks legislative and executive guidance. Initiatives are more likely to come from the bottom up—from local agencies and nongovernmental organizations—than from the top down, and the effectiveness of these initiatives is very much in doubt.

Modern water policy is a victim of post-Reagan minimalist government, which seeks to reduce the federal role in setting policy and increase the role of the states and local agencies. Today, there are no George Norrises, Robert Kerrs, Clinton Andersons, or Henry Jacksons—strong legislators who helped frame landmark natural resource-development projects and environmental policies, such as the Tennessee Valley Authority, the Arkansas River Navigation Project, and the National Environmental Policy Act—to articulate and implement visionary national and regional programs.

Instead of providing leadership, Congress is more likely to intervene in specific disputes to benefit a narrow but powerful local constituency, as it did when it recently forbade the Corps of Engineers from changing the Missouri River Master Manual, the federal guidelines that govern the river's management. One proposed change would have regulated the river's flow to improve habitat for endangered species such as the pallid sturgeon. The navigation industry and farmers, however, opposed any revision of the manual as it threatened to hurt them economically. President Clinton vetoed this meddling, and Congress was unable to override the veto. While this was a minor victory for river restoration, the process also illustrates how minimalist government will often lead to policy gridlock.

Third, the traditional mandate to allow for multiple uses of water resources had a firm legal foundation, unlike the new vision. The new paradigm—which attempts to balance environmental and economic interests—is contested by many traditional water users, and thus efforts to reform water policy can be blocked by powerful legislators beholden to them.

### *End of an Era*

The current minimalist federal water policy is apparent in two recent federal water-commission studies. Federal water commissions have historically played an important role in setting the national water agenda. In 1968, Congress created a National Water Commission to provide guidance for the expected continued large-scale regional water projects such as transbasin diversions, which were becoming increasingly controversial for both economic and environmental reasons. The commission's bold report examined all aspects of water policy, focusing on the use of markets to test the efficiency of alternative allocations. It also legitimized incorporating environmental values in water allocation and inadvertently provided a blueprint for the end of the reclamation era.

In 1992, in a similar attempt to review the complex mix of local, state, and federal water policy that no longer provided a clear vision of how to manage western water resources, Congress authorized the Western Water Policy Review Advisory Commission, appointed by President George Bush. This commission turned out to be a political orphan, however, after its sponsor, Senator Mark Hatfield of Oregon, retired from the Senate. The new Clinton administration tried to kill the commission, but Congress intervened to save it.

The strange mix of members on the commission—including private citizens, cabinet secretaries, and members of Congress—ensured that it was gridlocked from the start. Congressional republicans, for example, immediately took off the table the two most important issues: reorganization of congressional jurisdiction over water resources and the idea of executive integration of water resources activities. A final report—Water in the West: Challenge for the Next Century—was published after a majority of the commission reached a compromise with two members, each with a single idea. One wanted a watershed governance structure that would give states and local users more freedom to decide how to comply with federal environmental mandates, and the other wanted increased coordination of agency water resources budgets. The resulting report is like Richard Strauss' opera Ariadne auf Noxos, which overlays an Italian buffo opera over a Greek tragedy. The report endorses a controversial experimental watershed-management program that has the potential to displace federal standards with local ones. It also contains a full discussion of the transition from reclamation to reallocation, justifies a need for a new vision of river systems and watersheds, and examines the institutional implications of the end of the reclamation era. The report, however, seems to have suffered the fate of many similar reports and modern operas; they receive initial praise and interest and then disappear into the archives. As students of international relations would predict, the federal government's diminished role has set off intense competition among water claimants for supplies to meet the demands of urban growth and environmental restoration. In general, state governments have not taken up the slack and assumed the federal government s traditional mediating role. These developments place new stresses on water management and allocation law.

Environmental laws grant federal agencies considerable leverage to influence decisions about water allocation, but the regulatory agencies can neither make the necessary allocation decisions nor project operation changes. In addition, the regulatory power is often so drastic that the political costs of full enforcement are too high. Thus, the federal government Increasingly functions more as a facilitator of regional stakeholder settlements than as a regional development bank or traditional regulator, although the threat of federal regulation is always in the background.

### Conflict of Interest

Water law has two basic, related functions: to create correlative private property rights in scare resources and to impose limitations on private use, in the public interest. There is inevitable tension between these two functions. Correlative property rights recognize the rights of individuals to use a portion of a common supply while simultaneously protecting the interests of other users. Limitations on private use promote broader public values.

In general, states set the basic allocation rules, except for federal interests such as Indian Tribes and public lands. States also have the discretion to define the public interest, although they have not historically exercised this discretion to significantly limit private water use. State law continues to perform these functions, but it has remained relatively static. There are both virtues and vices in this stasis. The states have not been interested in taking up the slack of the ever-shrinking federal government. They primarily continue to administer the laws of allocation. On one level, this facilitates changes through water markets, which have emerged as a major reallocation policy instrument. Water rights have always been inalienable, but the prevailing assumption was that water would be used on the land where the right was initially applied. This assumption held more or less true when the principal form of conservation was supply augmentation, but this is no longer the case. Growing cities, power generators, and environmental interests are increasingly relying on the purchase of existing irrigation rights to meet the demand for new supplies.

The legal system's emphasis is shifting from setting the ground rules for the acquisition of rights and their enjoyment to lowering the transaction costs of transfers. Aside from specific commitments to environmental protection and restoration, there are few limits on the transport of water from water-sheds to cities outside the watersheds or on the power of cities to decide how much they need. But this limited role prevents the states from exercising control over the pace of reallocation or adjusting to new demands. For example, the insistence by many westerners that land and water are exclusive individual property rights with no community dimension undermines new community efforts to control their destiny. Land and water are alienable property rights, and individual rights holders are generally free to respond to market pressures without regard to the impact of their decision to break up a parcel of land or transfer a water right to the surrounding community.

### Goldrush Era

A recent decision by the California Supreme Court shows the danger of refusing to temper the protection of vested rights with accommodation of new demands. In 1996, a trial judge imposed a negotiated settlement on all groundwater users in the Mojave River Watershed. The decision did not strictly follow California groundwater law, but the state supreme court had a long tradition of approving allocation regimes that balanced groundwater conservation with equal access among all users. The court's decision that courts must determine the prior rights of all pumpers is, of course, a classic example of the rule of law, but it also illustrates the difficulties of adjusting, rather than eliminating, historic entitlements to respond to current conditions.

In an editorial, the Sacramento Bee characterized the opinion, somewhat inaccurately, as a "Gold Rush Era" decision, but the result shows that holdouts—chose who refused to accept the negotiated settlement—can use law to raise the costs of systemwide allocation adjustments that balance resource conservation and use.

The California decision is not an anomaly. In general, states have the poorest track record of incorporating environmental protection values into existing and future water-allocation regimes. There are some exceptions, of course. The Hawaii Supreme Court, for example, recently integrated the public trust doctrine, which provides that "public natural resources are held in trust for the benefit of the people," into its statutory water-allocation regime to impose potentially strict in-stream flow-protection duties on the state. Initially, the law of water rights has little weight to leave water in a stream, but more and more states are trying to maintain minimum stream flows to protect fisheries and other environmental values. In a dispute over the allocation of a ditch that carried water from the windward to the leeward side of Oahu, the Commission on Water Resource Management set interim in-stream flows for windward streams and denied several water-use permit applications. The court held that the commission c orrectly interpreted the public trust doctrine to give priority to the protection of fresh groundwater and surface-water resources because "the people...have elevated the public trust doctrine to the level of a constitutional mandate."

### Changing Tack

One possible side effect of the resistance of law to change is that courts may block reallocation schemes because they conflict with vested entitlements. This possibility has encouraged ad hoc solutions to conflicts, thus discouraging stakeholders from claiming their full entitlements. Around the country, all levels of government are trying a number of important ad hoc basin-restoration experiments to solve specific basin problems.

These efforts generally begin as an attempt to deal with a perceived crisis for threatened or endangered species. Once sufficient bipartisan political support has been mustered, however, they have evolved into a broader effort to accommodate both historic and new basin uses. The federal government remains an important participant, but power is shared much more broadly with states and stakeholders than in the past. The real question is whether these settlements will favour process over substance and in the end fail to reach long-term consensual solutions about how the real costs of reallocation will be shared.

The California-Bay Delta restoration process, for example, has been underway for almost a decade, but the parties have developed neither a clear focus—the restoration of the deteriorating Bay Delta ecosystem—nor a plan to accommodate ecosystem restoration with continued consumption for agriculture and municipal and industrial use. Many promising ideas such as adaptive ecosystem management—which sets ecosystem indicator targets and takes the necessary actions to meet them—have emerged, but the jury is out on their long-term success. The recent election and Supreme Court selection of a new president injects great uncertainty into the process and may encourage many stakeholders, such as irrigated agriculture, to dig in their heels and refuse to make necessary concessions.

### *Simulating Nature*

The recent federal legislation to fund restoration of the Everglades is an example of an effort to apply new management techniques to save a degraded ecosystem. The current thinking is that the system must be restored through intense management of the existing built system of canals and drains that caused its degradation. This is not a simple return to the status quo prior to development. Rather, it involves the artificial reconstruction of the environment before human intervention, using sophisticated techniques such as computer models of water flow and experimental management strategies that mimic the natural ecosystem.

Agricultural use of fertilizers containing phosphorus has contributed to unhealthy algal blooms in the Everglades. Experts widely agree that more low-phosphorus water must be put back in the system. In addition, diversion of water for agricultural irrigation has disrupted the natural cycle of water flow from north to south. Sheet flows must be more continual for longer periods of time during the wet season to sustain the glades during dry periods. Experimental releases of water into the glades have taken place, but the results are still uncertain as the experiments are conducted in the absence of scientific certainty about species and system responses to restoration efforts and management strategies. These efforts, therefore, must be constantly evaluated and often revised.

In February 1999, for example, a group of biodiversity experts complained to the secretary of the Interior that the federal government's actions had a high risk of failure because water releases into the park were insufficient to maintain the Everglades. Secretary Babbitt immediately agreed to the creation of a new scientific panel to monitor the experiment. In October 2000, the Senate approved a $1.4 billion-dollar restoration plan as part of a larger federal-state cost-sharing and cooperative strategy.

The Florida example may not be a good model for other restoration projects, however, because the Everglades are a heritage resource whose restoration has been widely accepted. In addition, federal and state governments are dealing with the major stakeholders—such as water-management districts, irrigators, and environmentalists—who will be adversely affected as more water is sent into the park and phosophorus loads must be reduced—in the old fashioned way: throwing money at them.

### *Doing it Right*

While American water managers grapple with an uncertain tangle of legal precedent and environmental manipulation, Australia is conducting an important experiment in flow maintenance and ecosystem restoration management on its largest river system, the Murray-Darling. The population of the Murray-Darling basin is relatively small, since Australia's population is concentrated along its coast. Still the basin contains 42 percent of Australia's agriculture—which consumes about 78 percent of the country's water supply—most of the country's major inland cities, and its capital, Canberra.

Like the United States, Australia is a federal system, and the Murray-Darling is an interstate river system. The Murray originates in the Snowy Mountains of New South Wales and Victoria, while the Darling originates in southern Queensland and joins the Murray near Mildura, Victoria. The system has been severely degraded—especially from increased salinity—due to diversions and dams. In 1992, the federal government and the basin states agreed on the Murray-Darling Initiative to conserve the river's ecosystem. The initiative led to the adoption of the federal-state Murray-Basin Agreement and the creation of a joint federal-state commission overseen by a federal-state ministerial council.

Unlike a United States interstate compact or an international treaty, the agreement imposes much more detailed land-use and water-management duties on the parties and is constantly being amended by new agreements. It has bite because it allocates the flow among the basin states, and it vests a commission with the power to control releases from specified upstream storage facilities. The Murray-

Darling Commission now runs the river—overseen by a ministerial council, composed of ministers from the participating territory and states—and a stakeholder advisory board composed of state representatives, farmers and other rural interests, environmentalists, and aboriginal representatives.

This experiment is the best available model for incorporating a river into a transboundary water resource managed to meet the needs of its many users. The effort has the three key elements that many of the current ad hoc United States experiments lack: a formal cooperative institutional structure, a relatively clear management objective, and a plan to limit inconsistent consumptive uses.

The most important precedent with potential international implications is the commissions adoption of a base-flow regime. This establishes the average quantity of water flow necessary to sustain a healthy ecosystem, and it mandates that any management regime should maintain that base flow. The commissions goal is to set base flows for ecosystem restoration, based on information about how flows affect the riverine environment. This regime is imposed by the law of the four basin states—Queensland, New South Wales, Victoria, and South Australia—on existing entitlement holders throughout the basin.

### *Cap and Trade*

The problem with establishing new regimes on developed river basins is that users have acquired, or at least claim to have acquired, vested rights. Yet federal and state governments recognize the need to limit water withdrawals, establish base flows, and stabilize and restore productive agricultural areas, especially those degraded by salinization. To that end, in 1996, the commission announced caps on water use.

The caps—which are the "cornerstone of a number of policies designed to manage water resources for scarcity: water trading, environmental flows and the security of property rights"—impose yearly limits on diversions of water in the four basin states and the Australian Capital Territory. Each state or territory's cap will vary from year to year according to the supply of water. The caps are administered by each state and will require aggressive management, since agricultural water diversions are increasing in both New South Wales and Queensland. In 1996 and 1997, three major subbasins in New South Wales exceeded the caps. Staying within the limits of the caps will require innovative management strategies, such as augmenting the surface-water supply with withdrawals of groundwater, abandoning the "use it or lose it" administration of water licenses, and implementing an accounting system to balance water use over a period of time.

Rollbacks in existing uses can be fair and efficient and at the same time promote environmental objectives. In major river systems, agricultural water use is almost always wasteful. Agriculture also uses more water than it is legally entitled to, so river managers have some flexibility to experiment with more-efficient use of agricultural water without unduly disrupting the expectations of legitimate users.

The most significant device Australia has used to ensure flexibility is the Pilot Interstate Trading Project in the Mallee Region of South Australia, Victoria, and New South Wales along the lower Murray River. Water prices and agricultural crops are comparable among the three states. Under the pilot program, individual diverters with high-security water rights such as irrigation licenses may sell water across state lines, provided that the water licensing authorities in each state agree to the transfer.

One of the major unresolved issues in water marketing is how to integrate the benefits of markets with environmental protection objectives. The Murray-Darling Pilot Program does this by establishing exchange values—the amount of water that can actually be transferred—among states. Trades by upstream diverters from New South Wales to Victoria and from Victoria to South Australia have a 1.0 exchange rate, which means that 100 percent of the entitlement can be transferred downstream. But transfers from South Australia to the upstream states of Victoria and New South Wales have an exchange rate of 0.9 so that only 90 percent of the entitlement can be transferred. Thus, the capacity of the lower river to continue to dilute the salinity will be protected. To integrate the program with the basin initiative, all transfers must meet a no-net-detriment-to-the-environment standard and must be consistent with environmental flows set for the Murray.

### Breaking the Gridlock

The current state of disarray in U.S. water-use policy leaves many critical interstate water basins dangling under Solomon's sword. While the Murray-Darling Initiative is still a work in progress, it could serve as a model for a new regime in water management in the United States. Either through presidential leadership or congressional initiative, policymakers need to forge a clear vision for the future, based on a common understanding that long-term environmental goals are also consistent with sustainable economic development. Otherwise, competing interests will drain the life from Americas rivers.

## Sustainable Land Resource Management: A Case Study in Chiang mai Province, Thailand

The rural communities in northern Thailand generally produce food using natural resources and practice a self-sufficient livelihood philosophy. Since the opening up of the Thai economy in the last four decades, agricultural modernisation oriented towards increased production using technology from the Green Revolution has been promoted by government agricultural polices and agricultural research projects. Adopting high-yield varieties, applying chemical fertilisers and pesticides and emphasising use of machinery are popular technologies supporting agricultural modernisation. From the viewpoint of sustainable development, however, there is a growing recognition that these new technologiee for increaeing production may reeult in environmental degradation. In fact, declines in eoil fertility and eoil texture, pollution of eurface/ground water, loss of biodiversity and food contamination are now becoming serious ecological issues, not only in Thailand but also in moet developing countriee in Aeia. Agriculture, on the one hand, interacting with climate, eoil, landform, water, foreete and biodiversity through production of crops and animals, and on the other hand, relating to farmers, rural communities, poverty and other social problems, plays an important role in ecological and economic eecurity of these countries. Farmers and the governments are therefore in a dilemma; they make every effort to enhance agricultural production under pressure from the market economy, but want sustainable management measures to maintain their land resources and environment and prevent rapid degradation (Keen 1983; Puginier 2005).

In the Ninth National Development Plan (2002-2006), the Thai government has oriented agriculture towards moderate and balanced development. The plan emphasises enriching farming knowledge and increasing production for sustainable agriculture and self-sufficiency according to local ability, improved farmland management and strengthening the commercial economy, especially for sustainable agriculture products. This chapter, mainly based on data from the agricultural sector of Chiang Mai Province, aims to identify the landuse factors benefiting sustainable land resource management in terms of environmental conservation. The conclusion and suggested measures should promote sustainable land management in both northern Thailand and other Asian developing countries.

## Agriculture Development in Chiang Mai

Chiang Mai Province is in the north of Thailand and has a total area of about 20107 $km^2$. Mountainous areas, over 500 m above sea level and covered with forests, take up about 80% of the land total and serve as the headwaters for many water sources, while agricultural and residential areas make up about 13% and 4% of the total area, respectively. According to socio-economic data for 2003, the population of Chiang Mai was 1.57×1 06. The gross provincial product was about $94.1 \times 10^9$ Baht, or 60x1 03 Baht per person. Agriculture is the third largest source of income and one of the most important economic sectors in Chiang Mai. Nearly 15% of the gross provincial product is from agriculture, and about 35% of the labour force of the province remains in this sector. Agriculture has developed rapidly in recent years in Chiang Mai. From the beginning of the Eighth National Development Plan (1997-2001) to 2003, the GDP for Thailand's agricultural sector increased by 5.5% per year, whereas the corresponding figure (the annual increase rate of the agricultural sector) during the same period in Chiang Mai was 8.5%. With the rapid development of agriculture, land-use change, both quantitative and qualitative, is apparent. Over recent decades, the agriculture area has increased from about $1.3 \times 10^5$ha to about $2.2 \times 10^5$ ha. Farmlands have expanded, even extending into forests in the mountainous area. Hill swiddens and fallows are cultivated more and more frequently. The fields and plantations have been aggregated in different ways and more high-value cash crops have been introduced from other regions and countriee.

Data and information collection have been conducted through vieiting the National Statistic Office of Thailand, field surveys, questioning local people and referring to publications. For evaluating the impacts of land use on the environment, four parameters, chemical fertiliser, chemical pesticide, land-use structure and land-use diversify, are taken into account. The land-use type in agriculture is categorised according to the cultivated products. The farmers' properties, income, land tenure and land area of the holdings in the agricultural sector were selected to recognise their influence on the four land use-related parameters. The analyses were conducted using statistical methods, where the dataset could satisfy such conditions, otherwise, qualitative and quantitative comparisons were carried out. A land holding is an economic unit of agricultural production (cultivating crops, rearing livestock or culturing freshwater crops) under single management, comprising all livestock kept and all land used wholly or partly for agricultural production, without regard to title or legal form. The holding's land may consist of one or more parcels, located in one or more separate areas of the same province.

### Increase in use of chemical Fertiliser and Pesticide

Although it is recognised that chemical fertiliser and pesticide are important causes of soil and water pollution and food contamination, more and more farmers have applied them to fields to increase agricultural production. In 1963, the number of holdings using chemical fertiliser was 7214, only 8% of the total holdings. In 2003, however, about 35% of the total holdings used chemical fertiliser, nearly eight times as many as in 1963. Farmland treated with chemical fertiliser reached 149209 ha, and about 48249 ton of chemical fertiliser (323 kg/ha) were consumed. The use of chemical pesticide showed a similar trend, with increases in the number of holdings and the percentage of holdings using chemical pesticides. In 2003, 75% of the total holdings reported using chemical pesticide. Unfortunately, the use of organic fertiliser and green manure has largely been ignored in this period. In 1963, the holdings using organic fertiliser were much higher than those using chemical fertiliser. In recent years, however, this figure has reversed.

### Change in Agricultural Land-use Structure and Diversity

Land use in agriculture in northern Thailand can be categorised into six main types: rice (including all paddy rice varieties), field crops (other crops except paddy rice), permanent crops (including all tree plantations, most of which are orchards), vegetable-herb-flower-ornamental plants, forest and pasture (consisting of planted trees and forage grasses) and other (including pen, freshwater culture and other small-scale cultivation types).

The comparieon of agricultural land-use structure between 1963 and 2003 indicates a significant decrease in rice and an obvious increase in permanent crops. In 1963, rice used 77% of the cultivated land, but in 2003 only 34%. In contrast, the percentage of permanent crops in the same period rose from 9% to 44%. In addition, the percentages of field crops and vegetable-herb-flower-ornamental plants slightly increased, and forest and pasture was reduced.

According to ecological theory, a diversified land-use structure could benefit the land and biological resource management, leading to sustainability. Therefore, the diversity of land-use patterns should be considered as an important parameter for evaluating sustainability of an agricultural system. In order to detect changes in land-use diversity during recent decades, the Shannon-Weaver index was calculated according to the land-use type and the composition of crops and other cultivated plants. The calculations show an increased tendency to land use diversity. In 1963, there were only 16 varieties of crops, including vegetables, and 12 species of fruit trees under large-scale cultivation, but in 2003, there were 28 varieties of crops, including vegetables, 13 varieties of flowers and 35 species of fruit trees. The change in land-use structure in the crop system may be related to the change in agriculture policy of the Thai government. From the First National Development Plan (1962-1966) to the Ninth Plan (2002-2006), national policy has gradually altered from first encouraging a strategy of growth-oriented agriculture, then a strategy of decentralisation and diversification, and finally a strategy of agricultural sustainability. Therefore, the agricultural land use has changed from mainly area expansion, through structural diversification, to etructural optimieation.

The Royal Project, launched in 1969, with the aim of offering hill tribes a viable alternative to the opium poppy through introducing temperate caeh crops, contributed to the increase in land-use diversity in Chiang Mai after 1978. Supported by the Royal Project, hundreds of fruit trees and vegetables species have been tested for their potential as cash crops in northern Thailand. Once theee fruits, vegetables and flowers were readily available to Thai consumers, local people also began to cultivate them on a large scale. In the Royal Project's orchards and gardens at Doi Ang Khang and Doi Inthanon, various fruit trees, flowers and vegetables from other regions and countries still grow today. During our field survey, introduced fruit trees were recorded in about 70% of the tribes and villages that we visited in the mountainous areas.

From the point of view of landscape ecology, diversifying crops, varieties and species on farming système is indispensable for reducing risks and uncertainty from a supply surplus or falling prices, as well as outbreaks of pests and diseases. The change in agricultural land-use structure in Chiang Mai has indicated, to certain extent, an improvement in sustainable agricultural land management through the diversified crop composition, which could benefit not only agriculture production but also the livelihoods of local people.

### Relationship Between Land use and Application of Chemical Fertiliser

To reveal the relationship between agricultural land use and the use of chemical fertilieer, two factors should be taken into account: the share of each land-use type to the total chemical fertiliser amount used in a year, and the amount used per unit area per year for each land-use type. The result of our analysis showed that permanent crops used the largest proportion of fertiliser in 2003: about 21299 ton, up to 44% of the total. This was followed by the vegetable-herb-flower-ornamental plants, at 12635 ton and 26%. Rice and field crops took the third and fourth places, respectively.

In terms of the annual amount per unit area in 2003, the vegetable-herb-flower-ornamental plants were first (673 kg/ha), followed by permanent crops (329 kg/ha). In comparison, the amounts for field crops and rice were only 272 and 200 kg/ha, respectively. Clearly the former two land-use types have played a very important role in raising the use of chemical fertiliser. In terms of agricultural sustainability, therefore, measures ehould be adopted to modify the variety composition in crop systems, as well as balancing the proportion of different types of land-use structure.

### Impacts of Holding size on Application of Chemical Fertiliser and Pesticide

Farm scale, annual income and land tenure of holdings were chosen to examine the relationship between the holding's size and farmer behaviour in the use of fertiliser and pesticide. Holdings in the agricultural sector of Thailand are usually divided into eight farm size groups and six income classes, according to the land area managed and annual income, respectively. The annual income differs between larger-scale farms and smaller farms. Most holdings had a land area lese than 0.8 ha (5 Rai) and earn less than 20,000 Baht per year, whereas the annual income of holdings with more than 6.4 ha (40 Rai) is above 10,0000 Baht. The land tenure of holdings has different forms: owned, rented, mortgaged, free and mixed. Among these, only the first two forms were involved in our analysis.

The amounts of chemical fertiliser used in 2003 by each farm group could be roughly characterised by a normal distribution between different groups, which is consistent with the distribution of total land area for each group. The larger the land area of a group, the bigger the amount of chemical fertiliser used by this group. At about 32%, the holdings with a land area between 1.6-3.04 ha (10-19 Rai) contributed most to the total fertiliser amount used by all of the farm groups. Analysis of the chemical fertiliser amount per unit area indicates, however, a different figure. The amount of chemical fertiliser used per unit area per year decreases with the increase in farming scale, from 479 kg/ha in the group with the smallest land area, to 201 kg/ha in that with the largest.

The use of pesticide differs somewhat from that of fertiliser. Taking the percentage of each holding group for pesticide and fertiliser use into account, the proportion using pesticide increases from the holding group with small-scale farms to those with large-scale farms, whereas the maximum proportion using chemical fertiliser was in the holding group with medium-scale farms over the laet two decades. In addition, it seems that the proportion of the holdings using pesticide has been increasing faster than that using chemical fertiliser over the past four decades. These data suggest that the large-scale farms might prefer to use pesticide in contrast to the small-scale farms.

For examining the impacts of income and land tenure on the use of chemical fertiliser and pesticide, the proportions of holding groups using chemical fertiliser and pesticide, the proportions of holdings in each income class, and those of holdings in each land tenure category have been calculated for each farm scale group.

The use of chemical fertiliser and of pesticide was influenced by properties of the different holdings. The impact of the holding's income on the use of chemical fertiliser was unclear from the correlation analysis. Among the six holding's income classes, only those of income class 20,000-50,000 Baht showed a significant correlation with the proportion of holdings using chemical fertiliser. Land tenure, however, seems to be a factor influencing the use of chemical fertiliser: the greater the land holding, the smaller the use of chemical fertiliser. A close positive correlation was observed between the proportions of the holdings renting land and the holdings using the fertiliser.

In contrast to the use of chemical fertiliser, the use of pesticide was often influenced by income instead of land tenure. The proportions of holdings in the lowest two income classes and those in the highest two correlate with the proportion of holdings using pesticide, negatively and positively, respectively. The foregoing analysis has indicated several landuse factors relating to the use of chemical fertiliser and pesticide, as well as to the diversity in cropping systems. Based on results of the analysis, some measures for land management can be suggested to improve the sustainability of agriculture.

### *Management of the Cropping System and Land-use Structure*

A diversification of the cropping system in Chiang Mai was clearly observed through our analysis, and has benefited both agricultural development and the conservation of the country's crop variety resources. The land-use change in cropping system had a positive influence on agricultural sustainability and increased farmer's income, balanced the geographically biased agricultural facilities, and enhanced agricultural resistance against natural disasters and market risks. However, the change in the cropping system also had some negative impacts on sustainable agricultural development, in terms of soil and water pollution, and food contamination. The shifting of crop structure, dominated by rice fields, to a more diversified structure of rice, high-value crops such as fruit trees and vegetables, together with the expansion of orchard area, has led to increased demand for chemical fertiliser and pesticide.

One of the most urgent tasks in promoting sustainable agriculture is to bring the use of chemical fertiliser and pesticide under control. Thus, strategic measures in land management for tackling the rapid increase in use of chemical fertiliser and peeticide should be coneidered and practiced. Our results suggest that diversification of the cropping system should be encouraged. But, at the same time, efforts should be made to encourage selection and introduction of new strains and varieties of crops and fruits with a lower demand for fertiliser and a higher resistance to diseases and insect pests. In addition, traditional soil improvement treatment, appropriate crop rotation, and other sustainable cultivation technologies should be encouraged in the management of the cropping system.

### *Balancing Large-scale and Small-scale Farming*

Large-and small-scale farming have different influences on the agricultural system in terms of sustainability. Practicing large-scale farming not only has the advantages of effective use of infrastructure and a high benefit-cost ratio in production, but also contributes to controlling the use of chemical fertiliser, since the amount of chemical fertiliser used per unit area is much lower in large-scale farming than that on small-scale farms. However, positive influences on sustainable land management have also been observed in small-scale farming. More small-scale holdings used less pesticide. Small-scale farming could also benefit diversification of the cropping system, due to a more flexible selection of crop varieties. Furthermore, it is easier for email-scale farms to practice traditional sustainable land management techniques, such as use of organic fertiliser, green manure and residue management techniques. For instance, the proportion of holdings using only organic fertiliser was about 15% in farms with less than 0.32 ha (2 Rai), whereas, in 2003, only 7.5% of farms with larger land areas used organic fertiliser.

In order to increase profits, farmers were encouraged to pursue aggregated production, which has led to a rapid growth of large-scale farming and an emphasis on mono-cropping. Considering the different impacts of large-ecale farming and small-scale farming on agricultural sustainability, balancing the farming scales in the agriculture sector should be taken into account in policy decision-making for future agricultural development. Ae for land management, epecial attention should be paid to containing the use of pesticide in large-scale farming, whereas the use of organic fertiliser and indigenous land improving practices should be especially encouraged among small-scale farmers.

### Insuring Land Tenure and Economic Incentives

Land tenure has been a major concern in terms of land sustainable management, because it contributes to a clear awareness of long-term land resource conservation. Our research found a negative correlation between owned land and use of chemical fertiliser. Quite often, farmers face a conflict of choice between long-term and short-term benefits. Moet sustainable land management practices pay attention to long-term objectives, while farmers are more concerned with short-term benefit for their survival. Therefore, ensuring land tenure with effective policies and legislative measures could encourage farmer's perception and adoption of sustainable land management technologies. If farmers do not own land, it is almost impossible for them willingly to make efforts towards long-term sustainable land management.

Although the ultimate goal of sustainable agriculture is food and environmental security and self-sufficiency for each farmer, the land management measures aimed at sustainable development show, in general, an increased demand for labour on the one hand, and a decline in competition potential for the products in domestic and international markets, on the other hand. Due to the coexistence of internal coste and external benefits in sustainable land management practices, economic incentives and other subsidies from governments and NGOs should be offered, both in the short and long term, to support the growth of sustainable land management. In policymaking, credit subsidies to farmers practicing sustainable land management, and tax reductions or remissions for inputs used in sustainable agriculture, as well as granting land tenure to communities or individuals should be considered. In terms of technology, training, information service, technical exchange networke and inputs to rural infrastructure construction should be enhanced to promote sustainable land resource management.

# Chapter 5

## Anthropology and Policy in Global Context

Just three days after the disbanding of the summit, anthropologists gathered at the University of Georgia's Botanic Gardens for two days of presentations at the Environment, Resources, and Sustainability Conference to discuss policy issues for the 21st century. Each participant presented work in the areas identified as priorities at the Johannesburg summit. However, no connection was made between our research and the United Nations debates. Consider what could be done with a systematic approach connecting the conference discussions to the global declarations of development need. An event of such high-level recognition as the Johannesburg summit could leverage bank and industry contributions and reinvigorate research partnerships among nonprofits, government, and educational institutions. Monitoring by researchers of policy events like the summit is critical to dissemination of research findings and formal input to future policy. The structure of the Athens conference included five working groups: conservation management and local knowledge, environmental justice, consumption and globalization, fisheries, and agriculture and water. Subgroups concerned with each of these areas attempted summary and prescriptive statements aimed at policy formulations toward sustainable futures.

Within the agriculture and water group, subthemes were: 1) contested uses of water; 2) maintaining clean water while sustaining agricultural and other uses; 3) allocation of access to water resources among users; and 4) management plans that honor the needs and perspectives of all users. All the discussions were profoundly anthropological in the sense that they began with individuals and their communities and sought ways to bring these viewpoints to the table in policy formation. At the end of the two days we had not just narrated the details of field projects and administrative struggles. We observed that as each of us argued for placing these projects on the policy agendas of relevant agencies or corporations, the term "policy" took on some of that magical quality that surrounds other iconic words in our discipline: "culture," "mental model," "rites of passage," and "ethnicity." But though we think that understanding culture grants us influence over decision making, most of us have little experience in policy work, particularly when it comes to multinational situations. However poignantly we researchers articulate hardships, injustices, or inhumane conditions and place them in holistic cultural contexts, it is the rare researcher who has the opportunity to take the next step: making findings valuable to those who could change the conditions that affect people and communities. The way to give life to research is to see it in action, not just in debate. Agencies, nongovernmental organizations (NGOs), and corporations typically have not looked to anthropologists for guidance in the formation of policy unless they have anthropologists on staff.

It was around these challenges that much discussion centred at the conference. Considering the global context of the Johannesburg conference, perhaps there is cause for some optimism that a bottom-up approach might be more successful in reaching goals of the many organizations intent on addressing water issues throughout the world.

## The Research-policy Cycle

To be considered for funding or included in policy discussions, usually, issues that researchers identify must be recognized by some politically and financially endowed body as worthy of placing on a policy agenda. The Johannesburg summit illustrated the cycle of policy formation and practice.

The first Summit on Sustainable Development in Rio de Janeiro in 1992 (the "Earth Summit") and the second in 2002 in Johannesburg reveal the high level recognition of global problems: conditions affecting one nation, or people, have direct and indirect impacts beyond immediate neighbours. For example, from the U.S. diplomatic perspective there is grave concern about political unrest fostered by unsustainable environmental conditions. Hence, the biggest budget contribution to sustainability projects is made through the U.S. Department of State.

A summit is not only an occasion for collaborative work among parties who otherwise work independently, it is also a means to name and institutionalize or politicize a problem. In the wake of these highly visible meetings and working groups, policy agendas have begun to emerge. Policy agendas are found within the United Nations, within and between nations, within NGOs, and also in a number of partnership agreements among private and public sector members as well as the NGOs. The U.S. Department of State Clean Water For People Initiative and the investment of $400 million by the United States to enhance water management worldwide are two examples of a high-level expression and support of policy. Once such policies with committed economic and political endowment exist, we can address anthropological involvement. What binds all of these agendas in principle is the political assertion by the United Nations that water is a high priority concern that must be addressed within and between nations for the physical health of the world's peoples, for peaceful relations among nations, and for the sustainability of the planet. Against that background an anthropological understanding of a community problem and the ability to recognize the genius of local systems of resource management are valuable.

Before we examine the contributions of the conference participants, we turn to two recent volumes, each focused on water and global issues, to delineate more precisely the predominant approaches expressed here. In their conclusion to a collected work on water access and use and power relationships, Donahue and Johnston (1998b:339-46) posit four dominant themes addressed by contributors to the volume. They order these as follows:

1. "Conflicts over rights and resources" in which power imbalances predominate. Local people's concerns are opposed to corporate and state agendas and practices; significant issues include local knowledge and sustainability.
2. "Project culture" that focuses on bureaucratic conceptualizations. This pits decision makers, usually official players, against those who are likely to experience the consequences of decisions and acts. This consideration also addresses "bureaucratic inertia," and interagency conflict, which is often the converse of meaningful action.
3. Development versus local sustainability of biological and cultural systems.
4. Crises with regard to water scarcity and quality, seen as cultural, social, and political issues.

The synopses presented here demonstrate the contributions of anthropological research and practice relating to the above thematic areas.

A second volume focuses on water issues relating to public health seen within a global context (Whiteford and Whiteford 2005). In their introductory statement, Linda and Scott Whiteford (2005:4) note that one of the "major threats to world stability" is "access to resources such as water." Anthropologists intent upon entering the discourse around this issue and becoming involved in practice addressing the crises, the need to engage in collaborative, interdisciplinary approaches, expanding our units of analysis to reflect the global milieu within which these crises materialize. This means taking account in our work of "international discourses, power structures, and commodity chains."

As Whiteford and Whiteford point out, international agencies are already significantly exploring approaches that have global reach. They note the implications, especially for local populations, of World Bank and World Health Organization policies that have come to privilege ideological perspectives like privatization. The oppositional response to some of these initiatives by local vulnerable populations has become the stuff of anthropological accounts, some of which are highlighted in this chapter.

Finally, with respect to anthropologists contributing to policy, Whiteford and Whiteford (2005:7) state: "the critical, structural approaches often employed in anthropology recognize that the social and political causes of structural violence against people based on gender, ethnicity, and economics require insertion into policy analysis." The contributions that anthropology can make to policy formulation may be obvious to us, but the enactment, as we see in the following cases, may be a more daunting task.

## Anthropology and the Research-policy Cycle

In the Sugar Creek Project directed by anthropologist Richard Moore (2002) of Ohio State University we find an illustration of the research-policy cycle. In this case a problem was identified by a federal agency report, thereby gaining local and national visibility: The Environmental Protection Agency (EPA) reported environmental conditions harmful to ecosystems as well as to human communities. Sugar Creek was ranked the second most impaired watershed in the state of Ohio, just behind the Cuyahoga River in Cleveland, which burned in 1969, just four years before the creation of the Clean Water

Act in 1973. EPA has developed the Total Maximum Daily Load (TMDL) plans for all impaired U.S. watersheds to reduce nonpoint-source pollution. The goal is for these TMDLs to be met by the year 2015.

Using the approach of participatory stakeholder involvement, whereby stakeholders play an active role in decision making, Moore created a research and action team that integrates stakeholders, specifically local farmers, with the interdisciplinary university research team. The research team consists of two anthropologists, a political scientist, an entomologist, a Geographical Information Systems (GIS) specialist, two water quality lab technicians, a Landsat specialist (NASA land satellite system), two natural-resources faculty specializing on riparian zones and microinvertibrates, a veterinary scientist specializing in E. coli pathogen transport, a USDA Agricultural Research Service hydrologist, two EPA research scientists, and one farmer. That team is part of a larger group of researchers that focuses on sustainable agriculture. The research team works together with 10 to 12 conventional farmers who came together to learn how to reduce nonpoint-source pollution, a reaction to Sugar Creek's TMDL top priority label. Moore's team also works with two other farm teams, which have addressed two other subwatersheds of Sugar Creek. In 2003 the team started working with people on local tributaries within the watershed, a major break from traditional agency policy.

Hackenberg (2002) recommends that projects identified for policy involvement possess three key criteria:

1. The project is being exploited within the private sector.
2. Proposals are prepared as participatory interventions.
3. Policies should engage social units capable of collective action to both produce and distribute benefits, which will provide improvements in quality of life as defined by the participants.

### *Moore's Watershed Project Meets These Three Criteria*

Field anthropologists engaging local farmers and other users of water regarding solutions to problems of allocation and supply may be cast in a different light from government policy makers, development economists, and functionaries representing international agencies. Often the anthropologist, operating through the lens of the village observer, may be criticized as simply myopic, harboring a perspective focused on the particular and incapable of grasping the larger picture. However, anthropology is positioned to identify problems by taking into consideration not only myriad factors that demand attention, but also the genius inherent in local systems of resource allocation and use where recognition is given to sustaining social and cultural orders.

Drawing on research in a number of local contexts, David Guillett addresses the new orthodoxy of demand allocation as a solution to an impending water crisis. From a point of departure suggested by Theodore Schulz and Robert Netting, he examines the institution of farmer managed irrigation systems (FMIS), nearly ubiquitous mechanisms for allocating water that take account of local conditions and imply sustainability. Guillett poses FMIS solutions against the market-based solutions increasingly promoted by international donor agencies, often as conditional contingencies of aid packages.

### Contested Interests and Their Resolution

Anthropologists have long done studies of small-scale farming in a variety of environments and under different regimes of land ownership, usage, and transfer. Fewer studies have been carried out with large-scale corporate agriculture or collective farming, although a classic study of singular importance was undertaken by Walter Goldschmidt in the mid-1940s (Goldschmidt 1978).

In addition to agricultural uses of water, papers considered a number of other topics, one of which involved the conflicting interests among various categories of users. Two of the papers dealt with contested issues between conservation and agriculture use and the attempts to mediate these. In the first of these, a planning process for uses in a small watershed area of southern Illinois was instituted. At a later date a team of researchers from Southern Illinois University examined the planning process. Their findings indicated that the process lacked local legitimacy and failed to be effective as a democratic form of resource management because it did not involve a widely representative body in the planning and it failed to establish institutions for local management of the watershed.

A second instance concerns the planning process and eventual management of a newly created national monument in Montana. In this case Douglas Midgett, interested in approaches to dispute resolution in conflict situations in the rural American West, followed the process through its early stages, paying attention to how local viewpoints were accommodated and stakeholders saw their interests represented in the development of the monument management plan. Midgett participated in the "scoping" process contributing to the formulation of the plan and has interviewed and advised both local parties and Bureau of Land Management (BLM) personnel, who will construct the final plan and manage the monument.

## Anthropologists as Participants in the Policy Arena

Diane Austin presented an example of an anthropological contribution to policy formulation at the University of Arizona where researchers and campus activists addressed water conservation and water use on campus. The work required "a thorough understanding of what is at stake for all interested parties as well as knowledge of the channels of communication, the social networks, and the issues under investigation." (Austin 2002:15). To make this expertise meaningful, she asserts that "we must examine and challenge our own biases and institutional policies that favour volume over substance, publication over application, exoticism over the familiar, and funding over need in project selection." (ibid., her emphasis)

Water resources and use are perennial concerns in the American Southwest. Although J. W. Powell recognized in the 19th century that aridity was the most important factor in human settlement in the West, boosters, planners, and policy makers have often avoided confronting this fact. A concerted conservation orientation has been shelved in favour of technological fixes, dam and reservoir construction, aqueducts that transport water hundreds of miles, and paper schemes that imagine even more fanciful "solutions." Within the Santa Cruz watershed that spans the border between southern Arizona and northern Sonora, Mexico, lie two major urban centres, Tucson and Ambos Nogales, the twin cities on either side of the border. Both urban areas depend for at least some of their water resources on the watershed, in-ground water, and limited surface flow. In examining two cases, Austin views attempts at a water conservation effort on the Arizona campus and describes a revegetation project in Nogales, Sonora, where forest destruction has contributed to desiccation and air-quality problems.

In this section we examine cases where anthropologists have become involved in policy issues at various stages, from advising and design to implementation on the ground, including direction of projects. We first look at a case where the anthropologist acted as a broker between conflicting interests and, in the effort to mediate the contentious issues, brought to the discussion an approach involving a proven dispute resolution procedure. In this case Inga Treitler was employed by the Oak Ridge National Laboratory to bring an anthropological imagination to the conversation, mediating between the Lummi Nation and hydroelectric development interests by illuminating cultural considerations that the Lummi people regarded as necessary to the discussion of environmental impacts. Her examination highlighted the cultural dissonances between the Lummi view of the resources, the position taken by the Federal Energy Regulatory Commission, and the inability of the environmental impact statement process to resolve these differences. A second case involved anthropologists working in a team to undertake evaluation of government-sponsored water policy initiatives. The team approach to examine the effects of water policy reforms in Zimbabwe included the anthropological contribution of William Derman and Anne Ferguson. Because a focus of the study was local access to water resources and empowerment, it was deemed necessary to locate the local within contexts of national policymaking and the agendas of international donor agencies. Applied aspects of the research centred on bridging efforts to link various levels in collaborative work. In this instance the issue of "project culture," alluded to above, became a significant consideration in how those policy reforms were carried out.

### Anthropological Research: Policy Implications

Finally, we present two contributions where the research of anthropologists, working alone, addressed issues having significant implications for policy. In one case the issue was local in scope, although it may have general application to a number of cases involving small-scale water-power generation. In the second, the issue was recognized as a fundamental problem of human exploitation of common resources, one encountered in a variety of historical and contemporary situations.

Richard Peterson's research addressed the cultural and social factors that influenced the successful implementation of development initiatives, in this case small-scale hydropower generators in tropical Africa. The natural resource in the form of permanent stream flows was abundant; the need for electricity to power machinery necessary to promote local productive activities was evident; and the projects seemed to fulfil development goals aimed at local enhancement and sustainability. However, as the research demonstrates, concepts of common property may collide with perceived individual ownership, a problem that can be overcome only through understandings of cultural factors, including powerful mythic figures whose agency in human affairs is taken for granted. In the Peruvian Andes, Paul Trawick examined irrigation systems long in use by agricultural communities. Again, communal principles with regard to water allocation were in operation. Trawick's analysis of these situations took a more nuanced approach to the "tragedy of the commons" discourse, for he informed us of the local structures that preclude the dire consequences of abuses of common property. In situations of water scarcity he argued that schemes for determining access must recognize the principles and privilege local ingenuity, especially when market based allocation is contemplated as an alternative.

## Water as Local Resource, Water as Global Resource

Although the issues examined in the works presented here illustrate the myriad problems emerging from the human uses of water, these by no means exhaust the list of issues that might concern anthropologists. When human manipulation of the land and its water resources is viewed in its entirety, this truism becomes abundantly clear. For example, none of the participants in the conference dealt with ground water or the problems of shrinking aquifers and implications for agricultural systems that depend upon these resources. Nor was the construction, and more recently, the removal of dams a topic our discussion. We know, however, of the dependence that states and communities have developed and of the disputes related to downstream flows that have surrounded discussions of the Missouri River system.

The scarcity of fresh water for human use worldwide has been tallied often: less than 1 percent in rivers, lakes, aquifers, plants, atmosphere, etc.; only one-hundredth of that available for drinking; and 40 percent of the world's population facing shortages daily (Johnston and Donahue 1998a:l). When we consider the other organisms with which we share the planet, the situation is equally dire: "fresh water ecosystems account for less than one percent of all habitat area on earth" (Postel and Richter 2003:26). Given this, we must approach the problems of water with the recognition that human beings are primarily responsible for the ecological degradation that has come to affect water quality and availability, and that it is only through human agency that these problems can be addressed.

In the epilogue to their volume on rivers and management, Postel and Richter (2003:202-204) suggest three actions that must be undertaken if we are to manage the resource in a sustainable fashion:

- Human communities must integrate better with natural cycles.
- We must reduce pressures on nature through reduced population growth and reduced consumption.
- We must address the implications of the worldwide trend toward commoditization and privatization.

Many of the contributions included here touch on one or more of these issues, indicating with careful study of local cases the validity of these prescriptions. In the conclusion to their edited volume, Donahue and Johnston (1998:339-341) discuss how the intersection of organizational forms produces conflict as competing interests, and, perhaps more importantly, different cultural concepts of water collide. We have seen this in a number of our contributions, and some of the papers discuss at length processes of dispute resolution and mitigation that have had mixed results.

One observation that seems inescapable is that anthropologists can contribute both to the compilation of knowledge through their research and to the formation of policy related to water management. Although we cannot blind ourselves to global forces that impinge on the communities and people we study, we need nonetheless to maintain a focus on the local and to assert the validity of our approach. Nitin Desai (2001:45), addressing the International Conference on Freshwater in Bonn, Germany, in 2001, invokes his "three Ps"-political will, practical steps, and partnership-a programmatic approach that aims to involve human beings at various levels in pragmatic and cooperative efforts to deal with these problems. And, as we noted at the outset of this chapter, Desai envisions the critical task as getting it right at the village level. If policy begins here, there is clearly a role for anthropology.

New market-based management strategies may prompt conservation and lead to higher-value uses of water resources. Freshwater is a scarce and often threatened resource throughout much of the United States, but particularly in the arid West. Supplies are being depleted or degraded by unsustainable rates of groundwater use, contamination, and damage to aquatic ecosystems. Meanwhile, demands for water are rising with an expanding population, higher incomes, and a growing appreciation for the services and amenities provided by streams, lakes, and other aquatic resources. But the options for increasing supplies are expensive relative to current water prices and are often environmentally damaging.

Providing for increasing water demands requires changes in how water has traditionally been managed and allocated among competing uses. There is a growing consensus that greater reliance on economic principles in managing and allocating water is critical for more efficient and sustainable use. More than a quarter century ago, the U.S. National Water Commission's Final Report to the President and to the Congress made a strong case for facilitating voluntary water transfers to promote a more-efficient allocation of scarce water resources and to curb the perceived need for additional water supply projects. In 1992, the International Conference on Water and the Environment in Dublin and the Earth Summit in Rio both endorsed viewing water as an economic good.

Introducing economic incentives was also one of the core recommendations of the World Bank policy paper on Water Resource Management prepared that same year. In 2000, the World Commission on Water for the 21st Century concluded that we are on a path toward a water crisis and that business as usual is unsustainable. The commission's proposals for changing course include recognizing that water is a scarce resource and that we need to manage it accordingly.

## Water Markets

Markets and prices play a role in allocation of resources among competing uses, and they provide incentives to conserve and invest in new supplies. In a competitive economy, price adjustments and market transfers keep supply and demand in balance. Prices rise when demand increases faster than supply. Higher prices provide incentives to use less, to produce more, and to develop and adopt technologies that conserve use and increase output. Markets enable resources to move from lower to higher-value uses as conditions change. For example, water traditionally used for irrigation may be more valuable as a municipal water source as the demands of a nearby urban centre increase. Tradable water rights potentially can encourage conservation and a more economically efficient allocation of scarce water resources. Currently, water is underpriced and often allocated based on institutions established when water was not considered to be a scarce resource. Users pay nothing for the water itself. Municipal and industrial users typically pay a fee reflecting the costs of storing, delivering, and treating water supplies.

But even these costs are likely to be subsidized for irrigation, which commonly represents a region's largest water use. Without an opportunity to sell unused supplies, irrigators have little incentive to conserve water. With the introduction of tradable water rights, however, users value water in terms of its opportunity cost—the value they could get by selling water—rather than at the subsidized price they pay for it. In spite of their potential benefits and growing popularity, market forces have been slow to adapt to the reality of water scarcity. Efficient markets require well-defined, transferable property rights, and the full costs and benefits of a transfer must be borne by the buyers and sellers. Both the nature of the resource and the institutions that manage and allocate water can make it difficult to meet these conditions.

### Obstacles

The variability of water supplies in time and space creates problems for establishing clear property rights. Driven by energy from the sun, water constantly evaporates from seas, lakes, and streams or transpires from plants, entering the atmosphere, then returning to earth through precipitation. Precipitation that is not quickly evaporated or transpired back to the atmosphere is the source of a region's renewable water supplies. This water flows into lakes, rivers, groundwater reservoirs, and eventually the ocean, unless it is first withdrawn for use.

Three basic systems—riparian rights, prior appropriation rights, and public permits—have developed for establishing rights to this water. Riparian rights. The common law system of riparian rights gives owners of the lands bordering a water body use of the water in ways that do not unduly inconvenience other riparian owners. Riparian rights have origins in the earliest legal systems establishing private ownership of land.

They continue to provide the basis of water law in many areas, including some arid Moslem and humid European countries and the eastern United States. These rights are poorly defined because shortages are shared by all riparian owners and use is subject to regulatory or judicial interpretation as to what is reasonable or might unduly inconvenience others. Moreover, riparian rights are not directly marketable because they are attached to the land and use is restricted to those lands.

### Appropriative Rights

Constraints on transferring water to non-riparian lands and uncertainties such as how much water a riparian owner can use are obstacles to applying riparian rights in areas where streams are fewer and flows are smaller and less reliable. Consequently, in the arid and semiarid western United States, riparian rights were abandoned in favour of water rights based on prior appropriation. Prior appropriation rights, which have been adapted by all 17 western states, have three principal features:

- Water rights are established by withdrawing water from its natural source and putting it to a beneficial use, such as irrigation. Unlike the owner of riparian rights, the party appropriating water does not have to be a riparian landowner, and the water does not have to be used on riparian lands.
- During periods of shortage, water is allocated according to the principle of "first in time, first in right." Thus, junior appropriators receive no water until senior appropriators—those with the oldest rights—have received their full allotment.
- Failure to use water for some period of time results in loss of the right. This provision creates a use-it-or-lose-it incentive, encouraging withdrawal even when the water contributes little if any value to the user.

Appropriative rights encourage the depletion of waters in a stream. Until recently, such instream flows lacked protection under state water laws. Unlike riparian rights, appropriative rights can be transferable, but sales are commonly restricted as to how and where water can be used.

### *Public Permits*

Riparian and appropriative water rights initially were acquired without state interference. But as supplies become scarcer, governments are assuming a more active role in controlling water use. Some form of a permit system now governs use of at least some water in virtually every country. In principle, water permits can be auctioned by governments and bought and sold in private markets. In practice, however, permits are usually free, and transfers are limited by the nature of the right as well as by the infrastructure available to store and transport water.

Transferring water from one place or use to another will commonly affect third parties—those other than the buyer and seller. When a farmer sells—in effect, transfers—water to a city, the economic base of the water-exporting community may decline. And when a transfer alters the quantity of water in a stream, other stream users are likely to be affected. Indeed, changes in flow affect the amenities and recreational opportunities that rivers and lakes provide, as well as the individuals who enjoy them. Generally, these water services are public goods that are not marketed, because the public cannot be excluded from freely enjoying them. Thus when marketing water, the private sector tends to ignore the impacts of water transfers on the public goods the waters produce. Similarly, polluters underinvest in waste reduction and treatment when the costs of using water bodies for waste disposal are borne by society rather than the individual polluter.

During the past three decades, water-related investments and legislation in the United States have been driven largely by a desire to protect the resource and the public benefits it provides. The resulting environmental legislation and regulations have contributed to uncertainties over water rights. On one side are the traditional users with rights established when water was treated as a free resource and environmental impacts were ignored. On the other side are the more-recently empowered stakeholders armed with legislation designed to protect and restore environmental and recreational uses.

While domestic, industrial, and agricultural users may compete for the water diverted from streams and reservoirs, all three groups vie with environmentalists and recreationists over the amount that can be withdrawn. Conflicts also arise over the priority that dam operators give to flood control, water supplies, hydropower production, fish habitat, and recreational opportunities. Without markets and prices to guide allocation of water and guide dam and reservoir management, conflicts are often played out in the courts or administrative proceedings.

Groundwater initially was treated as a resource that landowners could capture at will. But groundwater is often a common property resource that flows from one property to another until captured for use. Pumping can adversely affect third parties. One party's pumping can reduce the water available to neighboring water users, forcing them to pump from greater depths and lowering their well yields. Third parties may also be harmed if groundwater use reduces surface flows, causes saltwater intrusion into an aquifer, or results in the collapse of lands above underground aquifers that have been at least partially drained. The emergence of such impacts, along with improved knowledge of the links between ground and surface water, has led to restrictions on groundwater use. In the western United States, for instance, most states have adopted some form of a permit system for groundwater. But even in states lacking a permit system—such as Texas, which continues to grant landowners unrestricted rights to pump groundwater without liability for damages inflicted on others—landowners overlying a defined aquifer may voluntarily form a conservation district to regulate wells.

### *Opportunities*

The absence of markets and market-based prices to allocate scarce supplies and guide water managers has resulted in large differences in the value of water among alternative uses. For example, most of the water rights in the western United States are held by farmers and irrigation districts, which pay only the modest cost of having water delivered to their farms. As a result, large quantities of water are applied liberally to relatively low-value crops. In some cases, simply leaving more water in the river to provide hydropower, recreation, and fish and wildlife habitat might increase the total value of the water to society. Similarly, selling water to urban areas that otherwise would invest in costly and often environmentally damaging water-supply projects might boost water values by an order of magnitude or more in some cases. Large differences in the value of water among alternative uses provide powerful incentives to overcome obstacles to transfers. As a result, water marketing is becoming common and increasingly innovative in several countries. A variety of market arrangements have emerged in the western United States to accommodate and respond to short-term fluctuations in supply and demand stemming from climate variability or other factors. These include leases, options to purchase water during dry periods, and water banking. The temporary nature of such transfers blunts a principal third-party concern that a transfer will undermine the economic and social viability of the water-exporting area.

Farmers with senior appropriative rights who grow annual crops might profit by selling an option to use some of their water during a drought and leaving some or all of their fields fallow. Growers threatened with the loss of long-term investments in orchards, or cities facing rationing, are able to more than compensate these farmers for losses incurred from temporarily leaving some of their lands fallow. Such transfers among farmers within the same irrigation district are common and often relatively easy to arrange. But cities seeking to transfer water away from an irrigation district are likely to encounter greater institutional obstacles and financial costs.

## Water Banks

Water banks represent one solution to these obstacles. Water banks are designed to facilitate water transfers in response to short-term changes in supply and demand conditions. They enable the owner of a permanent water right to sell all or part of one year's entitlement. Thus, the entitlement is, in effect, leased or rented but not permanently transferred. The primary objective of a water bank is to bring together those who want to purchase water with those who are interested in selling their entitlements.

The bank provides several key functions. It determines which water use entitlements may be banked and the quantity of water associated with each entitlement. It determines who can rent or buy water, and it sets the rental rules. The bank may also determine how much water can be transferred without injury to third parties. Once the rules are established, the bank operates like a broker, accepting valid water use entitlements for deposit and making them available to those hoping to obtain water at a lower cost than they would otherwise have to pay.

***California :*** A temporary federal water bank administered by the Bureau of Reclamation and endowed with federal funds was established in California during the 1976-77 drought to provide for water transfers within the agricultural sector. Then, in 1991, California established an emergency water bank in the fifth year of a prolonged drought, after legislative attempts to promote private transfers produced few transactions. The bank, which reallocated water among willing buyers and sellers, operated in 1991, 1992, and 1994. Initially, the bank purchased water for $125 per acre-foot (about 10.1 cents per cubic meter) and sold it for $175 an acre-foot plus delivery charges. In some cases, the delivery charges exceeded the initial cost of the water. Purchases by the bank proceeded slowly until the state guaranteed sellers that their price would be adjusted upward to reflect subsequent seasonal price increases. Farmers willing to idle land or shift from diverting surface water to pumping groundwater were the principal sellers. The bank purchased 800,000 acre-feet (nearly 1 billion cubic meters) of water, based on early estimates of critical need before the buyers made firm commitments.

California incurred a sizable financial loss when the bank was able to resell only about half of the water. Unsold water was used primarily for carryover storage and for reducing saltwater intrusion into the delta of the Sacramento and San Joaquin Rivers. In subsequent years, the state required a signed contract with a buyer before committing to purchase water. The third-party impacts associated with these transfers are unknown, but they were probably insignificant compared with the benefits of moving water to higher-value uses during a period of severe drought.

***Idaho :*** Idaho established the nation s first permanent water-banking program in 1979, and several western states are considering similar actions. Idaho's bank allows temporary or permanent transfers of water rights, but rules governing where and how the water can be used are restrictive. For instance, out-of-state transfers are prohibited and irrigators receive preference over all other users in purchasing or renting banked water.

***Australia :*** In Australia, the state of Victoria, which facilitates transfers, defines water rights as explicit shares of stored, rather than delivered, water. Under this system of capacity sharing, decisions regarding reservoir releases are made by individual owners of the rights rather than by a central authority. Reservoir operators serve like bankers, making releases on request. The operators also keep track of each owner's balance on a continuous basis by adding inflows and deducting releases and losses from evaporation and seepage. Water users control the timing of their deliveries, and transfers can be made simply by having the operator make the appropriate debit and credit.

Banks can operate at any administrative level, ranging from multi-state to water districts to ditch companies. They can be designed to manage different types of water-use entitlements. And they can facilitate temporary water transfers by developing clear, well-defined rules and procedures that reduce transaction costs. Temporary water transfers, however, are not particularly effective for adapting to the long-term demand and supply shifts that result from population and income growth, urbanization, rising values for instream flows, groundwater depletion, and climate change. Indeed, at some point, as supply and demand conditions change, the historical allocation of water rights becomes inefficient enough to warrant a permanent transfer of rights.

### Keeping Deserts Green

Transfers of permanent water rights are permitted, subject to review of third-party considerations, in all the western U.S. states.

The process of resolving third-party impacts is often slow, costly, and contentious, however, and the outcome of a proposed transfer can be uncertain. Those orchestrating the proposed transfers face either the challenge of proving that a change will not harm others or the added cost of compensating the third parties who might be adversely affected by the transfer.

Ongoing efforts to meet the water demands of the rapidly growing coastal area of southern California illustrate the challenges of securing additional water in a region where supplies are already fully developed and allocated. The task is made more urgent and difficult because access to some of the region's traditional sources has been blocked. As a result of environmental concerns, for instance, Los Angeles has been forced to reduce the water it takes from Mono Lake and Owens Valley. The region is also losing rights to the unused entitlements of other states to water from the Colorado River.

The Imperial Irrigation District (IID), in the southeastern corner of California, owns senior rights to much of California's share of the Colorado River. To help meet the growing demands of the state's southern coastal region, the Metropolitan Water District of Southern California (MWD) agreed in 1989 to invest approximately $115 million, plus about $3 million annually for operation and maintenance, to conserve water in the irrigation district through such measures as lining canals to prevent water loss through seepage. In exchange, the MWD acquired the rights to about 100,000 acre-feet of conserved water per year. More recently, San Diego—which depends on, but has a low priority claim to, MWD water supplies—agreed to fund additional conservation investments in the IID in return for conserved water.

These agreements illustrate the opportunities and obstacles associated with using market forces to reallocate water. In these cases, water was transferred from relatively inefficient, low-value agricultural uses to higher-value urban uses; IID's agricultural base was preserved through conservation investments; and the interests of neighboring U.S. irrigation districts were protected. Yet concluding these apparent win-win arrangements required nearly five years of often contentious negotiations among the participants and interested third parties.

Moreover, agreement in this instance was facilitated because the participants were able to ignore the adverse third-party impacts in Mexico. Indeed, before the canals were lined, some of the water seeping out of them had helped recharge groundwater aquifers used by Mexican farmers just across the international border. These third-party impacts were ignored because the Mexicans had no recognized legal claim to the water. Efforts to arrange interstate sales of Colorado River water have been less successful. The 1922 Colorado River Compact among seven western states divided the river equally between the upper basin states (Colorado, New Mexico, Utah, and Wyoming) and the lower basin states (Arizona, California, and Nevada). The upper basin states have never fully used their entitlements. The MWD therefore has been able to take the unused water for free. California, however, is being required to cut back on its use of these surplus flows, and the upper basin states are seeking opportunities to benefit from their full entitlements.

The potential for mutually profitable transfers from the underused upper to the overused lower basin has stimulated several proposed sales. In the 1980s, the Galloway Group, a Colorado corporation with claims to 1.3 million acre-feet of Colorado River water, proposed constructing reservoirs to produce hydropower and store water for leasing to Arizona and southern California. San Diego paid $10,000 for an option to lease 300,000 to 500,000 acre-feet per year for 40 years, but the project died under a flood of unresolved legal issues.

Las Vegas—which already uses most of Nevada's entitlement to the Colorado River—is currently seeking more water. Meanwhile, it lacks rights to surplus flows, and depletion of its groundwater is causing subsidence within the city. Unused upper-basin entitlements to the Colorado River are a logical source of additional supply, and Utah appears to be a willing seller. But consummating this or any other water transfer between the two basins could require renegotiation of the 1922 compact. Transfers among the lower basin states encounter fewer legal hurdles but, because of their similar hydrology, offer fewer economic benefits than do transfers between the upper and lower basins. In 1997, the U.S. Department of Interior issued a ruling designed to encourage and facilitate voluntary transactions among the three lower-basin states. Arizona has established a Water Banking Authority to purchase its own unused entitlements for storage in groundwater basins and possible sale to California and Nevada. But the opportunities for profitable water transfers among these three states—each of which is trying to meet the demands of rapidly growing metropolitan areas—pale in comparison to the potential benefits of transfers to the lower basin from the upper basin, with its large quantities of unused entitlements.

### *Colorado-big Thompson*

The U.S. Bureau of Reclamation's Colorado-Big Thompson project has been cited as a prime example of efficient water marketing. The project involves a series of reservoirs to capture part of the flow of the Colorado River and its tributaries. An average of 230,000 acre-feet of water annually is transferred through a tunnel from the western slopes of the Rocky Mountains to the Northern Colorado Water Conservancy District in northeastern Colorado. Rights to proportional shares of this water are freely traded, unencumbered by third-party concerns, within the district.

Under western law, downstream users generally own rights to the return flows of upstream users, and transfers must take account of downstream impacts.

But since Colorado-Big Thompson water originates in another basin, the district has retained ownership of the return flow of the diverted water. This arrangement does not eliminate third-party impacts, but it does eliminate the need to consider them in transfer decisions. In this case, the benefits of being able to transfer water readily among agricultural, municipal, and industrial users within the district are likely much greater than the costs of ignoring the third-party impacts. But limiting sales to within the conservancy district precludes even more-profitable transactions that might take place with buyers outside the Northern Colorado Water Conservancy District. For example, the right to an acre-foot of water in perpetuity has sold for $3,500 more in the neighboring Denver suburbs than in the conservancy district.

### *Chile Waters*

Chile, which has embraced market economy principles for the last quarter century, has introduced measures to encourage water marketing. In 1981, the nation separated water rights from land ownership and made these rights freely transferable. Views differ as to the extent and benefits of the water markets that emerged in response to these changes. One view holds that Chile's water markets function effectively; water moves from lower to higher value uses, prices are responsive to temporary as well as longer-term scarcity, and trading is active.

A less-optimistic view holds that transfers of water rights separate from land ownership are uncommon, involve only a small percentage of users, and result in little actual reallocation of supplies. In particular, several factors are cited as inhibiting sales of water rights separate from land ownership. These include the inflexibility of the existing canal systems for distributing water, uncertainty as to who actually owns the water rights, a rural culture that believes water should not be bought and sold separately from land, and slow and erratic administrative procedures designed to protect third-parties from injury.

## Sustainable Management

A consensus is growing that sustainable economic development depends on treating water as a scarce resource and using economic principles to guide its management and allocation. Water markets are a means of introducing these principles and allocating supplies in response to changing supply-and-demand conditions. But marketing water differs in important ways from the sales of most goods and services. The fugitive nature of the resource, the variety of services it provides, and interdependence among users limit the potential for efficient water marketing. Additional constraints result from the laws, regulations, and treaties that establish rights to water and limit how it can be used.

Water resources within a basin—precipitation, runoff, water in lakes and streams, and groundwater—are interrelated. Water users become increasingly interdependent as supplies become scarcer. Dams, reservoirs, canals, pumps, and levees make water availability less dependent on the vicissitudes of the hydrologic cycle and more dependent on human decisions. This infrastructure broadens the opportunities for allocating supplies and generates new demands on the resource. Reservoir operations affect a variety of water uses, such as flood control, hydropower production, recreation, fish and wildlife habitat, navigation, water quality, and domestic, industrial, and agricultural water supplies. Allocating reservoir capacity for one use affects other users within the hydrologic system. Drawing down a reservoir for flood control, for example, may reduce available supplies when they are most valued for irrigation, hydropower production, navigation, or recreation.

Managing these reservoirs in a way that best serves society is a daunting challenge. In smaller river valleys where water uses are limited, capacity sharing, as practiced in Victoria, Australia, is a way of defining marketable water rights. In a larger, more complex system, the interdependence among users is too great to ignore. Market allocation of water or reservoir capacity in such a system would be inefficient and chaotic. While markets are not a panacea for achieving efficient and sustainable water use, they can play an important role in achieving these goals under some circumstances. Water markets, whether formal or informal, have a long history of facilitating transfers. Chile and the western United States, two areas where marketing is most advanced, illustrate both the opportunities and limitations of water markets. Water marketing in these countries has largely involved transfers from relatively low-value, inefficient irrigation use to higher-value domestic and industrial uses. Moreover, the sales often provide incentives and funds for water-conserving investments to protect the economic base of the water-exporting community.

Geography and institutional factors, however, have restricted development of water markets. Moving water outside of its natural channels is costly and subject to economies of scale. Chile's rivers flow from the Andes to the ocean in a series of small, steep-gradient rivers separated by hills. Consequently, it is expensive to move water from one watershed to another or from downstream to upstream areas within the same basin. The frustrated efforts to sell water rights from the upper to the lower Colorado River basins illustrate how institutional factors can limit potentially profitable transfers even where the infrastructure is in place to move water at low cost. The successes and failures in transferring water emphasize that clearly defined, transferable rights are a necessary, but not sufficient, condition for market transactions. Once transferable rights are established, the most important challenge for creating efficient water markets is developing procedures for expeditiously and fairly handling third-party impacts. Unfortunately, these impacts are not always obvious or quantifiable. Both Chile and the United States allow transfers subject to consideration of these impacts. But the judicial and administrative procedures used to assess these impacts and compensate third parties often impose a high hurdle for prospective buyers.

In spite of the obstacles, the potential gains of transferring water for new uses are encouraging the development of water marketing in many areas. The incentives for voluntary water transfers are strong and will continue to grow as the resource becomes scarcer and the costs of providing water for traditional uses increase.

### A Hot Future

Global warming would likely add to the potential benefits of water transfers. A warming world would alter the hydrologic system and increase the demand for water. The magnitude, timing, and even direction of climate-induced changes in a region's water supplies are uncertain. The costs of building dams, reservoirs, and canals in anticipation of these uncertain changes are high. But reexamining reservoir operating rules, relaxing constraints on water use, and developing institutions to encourage voluntary exchanges of water through markets would create a system more efficient and able to adapt to whatever the future might bring.

Will we run out of fresh water in the 21st century? The media highlights the parched lands, dry riverbeds and springs and falling groundwater tables across the world daily. Over a billion people living in developing countries without access to safe drinking water are facing economic and water poverty. Another real and troubling indicator is the rapid rate of aquatic habitat degradation and biodiversity loss in the last century. Projected changes in climate due to greenhouse gases invariably portray a future world that is much drier in the tropics—where over half the world's population lives—and suggest a global increase in floods and droughts.

Is a global water crisis already upon us? The answer to this question seems to depend on who you ask. On the one hand, active voices such as Sandra Postel, Peter Gleick, Vandana Shiva, Lester Brown and Paul Elrich, as well as leaders of major global organizations with an interest in water, have been warning of an impending global water catastrophe. On the other hand, the mainstream academic community involved in hydrology and water has largely ignored the topic. For example, a Google search for "water crisis" leads to almost 1 million hits, but the same search on Google Scholar yields approximately 4,000 hits as compared to over 1 million Google Scholar hits for "climate change." Malay of these articles focus on policy solutions, but do not necessarily explore the nature of the problem in-depth. Furthermore, the literature is largely non-American and contains references to much of the same work. Introducing "global water crisis" into a Google search reduces the number of hits by a factor of ten. In fact, the handful of scientists who do study this problem have divergent opinions as to whether and when the world will run out of water. A handful of scholars—particularly economists—go so far as to claim that a global water crisis does not exist or is, at best, overstated. These scholars generally find that, on the whole, water access is improving worldwide and that with continued efficiency enhancements, the amount of water will continue to meet existing demands. Perhaps the way the global water crisis has been defined—whether the world will run out of freshwater—is the wrong way to look at the problem. While there are many scholars looking at the range of localized and specific water challenges that are occurring around the globe, it seems that the academic community has yet to find success in accurately characterizing the sum of their parts. In this chapter, we argue that there are three distinct water crises—or challenges, depending on who you ask-that have yet to be systematically connected by scholars. It is by looking at how these three challenges are interrelated that we can better articulate the global characteristics of water resource dilemmas and, ultimately, identify the global factors that can help solve these dilemmas.

## Reorienting the Debate: Three Crises Rolled into One

Three types of water crises appear prominently in academic and professional discourse. First, there is the crisis of access to safe drinking water. This includes the inability to provide basic infrastructure to store, treat and deliver water supplies to a large part of the world's population. Second, there is the crisis of pollution that is analogous to climate change in that it relates to the impact of by-products of resource use. Third, there is the crisis of scarcity, or resource depletion, which is analogous to the fear of running out of oil. Now that we have defined three types of water crises, we can examine what we know about them, how they are linked, to what extent they are global problems and, finally, what are some possible solutions.

### *The Access Crisis*

Many people equate the global component of a water crisis with the vast number of people worldwide whose economic productivity and social development is limited by access to safe drinking water. For instance, the World Health Organization, the World Bank Group Development Education Program, Global Water and the Global Water Challenge draw attention to the fact that over 1 billion people lack access to safe drinking water. As a result, the United Nations Millennium Development Goals, the World Water Forum and other groups have rallied around a common metric for this issue by measuring the number of people with access to safe drinking water. For example, one of the key targets under the Millennium Development Goals is to "reduce by half the proportion of people without sustainable access to safe drinking water." Although these goals have been lauded as important policy directives, the international community has not yet made much progress in meeting them. Why is it so difficult to meet these goals? A vast body of literature points to the technical, institutional and financial challenges involved in developing the infrastructure and systems needed for water storage, supply and treatment. As this body of literature has discussed, poor countries may not have access to sufficient capital to build large-scale infrastructure like reservoirs, water treatment plants or delivery systems. If they have donors to supply the initial capital, they often do not have the means to repay these loans. Some developing regions that acquire the resources to build new infrastructure later discover that they cannot afford to maintain it. Other times, donors build water supply projects that are grossly mismatched with the needs of local communities.

Despite these challenges, research and practical experience have shown that the water access problem can be addressed—at least superficially—with existing knowledge and resources. For example, numerous low-cost technologies are now available to treat water quickly without large-scale infrastructure. Simple and cost-effective infrastructure, like rainwater harvesting systems, is readily available to many water-scarce communities. Additionally, creative financing mechanisms—like public-private partnerships—have been adopted by many local communities to pay for new water supply systems. These types of interventions—as well as successfully developed large-scale water supply infrastructure—have helped increase the number of people with nominal access to safe drinking water from 77 percent in 1990 to 83 percent in 2002. It is because of these improvements in recent decades that some scholars have argued that a global water crisis does not exist.

### *The Pollution Crisis*

Intertwined with the access crisis is the water pollution crisis. For those 1.1 billion individuals who lack access to safe drinking water, "safe" is often the key word. While the infrastructure for water storage and access is often available, sometimes the water is contaminated by chemicals, microbes or other pollutants that render it non-potable. Yet, just as we have the know-how to develop water supply technologies, we also have the know-how to treat contaminated water. In the last century, tremendous research efforts have translated into an ability to treat wastewater and remove many of the most well-known chemicals of concern, including the ability to reuse or release this treated water into the environment. Technological advancements and tighter environmental regulations have created major progress in controlling the pollution of water from point sources, such as industries and municipalities. Although exotic or emerging chemicals continue to be a concern, their control is an active area of research. Therefore, where there is political will and available funds, pollution emitted from point sources is now under control. However, the political will and funds available to control point sources are still limited in many regions of the world. Many poor countries face similar challenges in developing the infrastructure to treat water as well as in supplying water.

An even more difficult pollution problem to solve is that of non-point source pollution, which results from diffuse sources, such as farms, or is caused by atmospheric deposition from industrial polluters. This form of pollution can have wide-ranging impacts, both on human use and on ecology, particularly through the accumulation of contaminants in water bodies and through the biological food chain. Examples of large-scale and cumulative ecological effects include hypoxia in the Gulf of Mexico, pfiesteria in the Chesapeake Bay and the decimation of the Ganges River dolphins. Historically, the effects of non-point source pollution have been easier to ignore than point sources because they affect humans less directly and visibly than sludge coming out of a pipe and directly polluting a drinking water source. Often, the cumulative impacts of non-point source pollution do not show up until they harm habitat and species living in downstream estuaries, bays and wetlands. These impacts cannot be ignored forever. As China has recently discovered, the extensive pollutants entering its waterways from factory waste, agricultural runoff and municipal sewage have had a tremendous impact on the quality of their aquaculture, causing decreased international confidence in their seafood markets.

Although it is possible to reduce pollution from diffuse sources, it is typically challenging and costly First, limiting non-point pollution requires substantial time and effort to figure out from whom and where the pollution is coming from, especially when the total amount of contaminants is high and when large volumes of water are moved during rainfall. In any given watershed, we may generally know that nitrogen and phosphorous entering a river is coming from upstream farms or mercury is coming from deposition produced by regional power plants. However, it is often quite difficult—without direct and costly monitoring—to know how much each polluter contributes in a given location. Thus, this type of pollution is more difficult to regulate and control than point sources. The United States has struggled for decades to enforce the Total Maximum Daily Load (TMDL) requirements under the 1972 Clean Water Act. These requirements call for states and the federal government to identify sources of pollutants for each of the nation's waterways and to set acceptable limits for each· pollutant.

One approach to dealing with challenges of identifying non-point source polluters is to require all industries that produce diffuse pollution to adopt technologies or practices that reduce the flow of pollutants into waterways. Many of these technologies are known and relatively simple to adopt. For example, all farmers could plant riparian buffers to filter nutrients or cover crops to reduce leaching. Alternatively, ranchers could protect streams using fencing to keep livestock fecal matter from entering waterways. The problem with these solutions is often political.

If we examine the experiences of some of the wealthiest regions in the world, like the Chesapeake Bay watershed—which encompasses Washington, DC and six U.S. states—large industry polluters often vigorously oppose such regulatory actions because they affect their bottom line. Enforcement or consensual action is often hampered by the fact that there is very limited data on the effectiveness of these best management practices, and a cost-benefit analysis is therefore difficult.

Moreover, unlike point sources of pollution—where it is possible to estimate the quantities of pollutants in a watershed and the benefits of reducing those pollutants—predicting the effects of non-point source pollution is challenging. For example, stochastic extreme meteorological events can lead to sporadically large "loadings" of non-point source contaminants, which are difficult to predict. The cumulative impacts of a series of such events over a long period of time, and over large distances, can make it even more difficult to understand and estimate these pollutants. Mechanisms of non-point source pollution may also entail transport through multiple media. For example, this may include volatilization into the atmosphere, followed by deposition at a different location through rainfall, followed by the binding of pollutants into riverine or lake sediments. These mechanisms can operate from local to regional to global scales. Consequently, cause-effect analysis and monitoring for compliance or scientific analysis are made considerably more difficult.

Various incentives, such as low-cost loans or tax breaks to encourage voluntary pollution reduction measures among industries, can also fail. It is difficult to monitor and enforce non-point source pollution standards in large watersheds with multiple and diffuse polluters. Often, the opportunity looms for any one polluter to catch a free ride off the efforts of others. Moreover, when non-point source pollution crosses political boundaries, upstream states have little incentive to control or treat pollution for the benefit of downstream states—unless downstream states have substantial political or economic bargaining power. Thus, non-point source pollution remains a significant challenge for technical, as well as socioeconomic and political, considerations.

### *The Scarcity Crisis*

The third water crisis is one of scarcity. Scarcity refers to a situation when the water supply is inadequate in relation to the water demand for basic human and ecological necessities, including the production of food and other economic goods. Scarcity is arguably the principle component to the threefold water crisis because scarcity can drive—or at least exacerbate—both water access and water pollution. A community that has pumped all of its shallow groundwater dry will find it much more expensive to build deeper wells or to build trans-basin diversions to bring surface water in from another region. Additionally, when water supplies are depleted from a watercourse, the pollutants that may have accumulated there over time are likely to be more concentrated, thereby exacerbating the pollution crisis. This connection between water scarcity and pollution can then, in turn, lead to problems of water access. For example, municipal water supply systems can face increased treatment costs when reservoirs or instream flows are low because pollutant concentrations increase in the water they must treat. Water scarcity not only makes it more difficult to get adequate and clean water to meet human needs, but also harms aquatic habitats and species downstream.

With regard to access, invariably the most economical sources are developed first. Therefore, as scarcity increases, the level and reliability of access to water suffers unless water system budgets are increased. As such, we implicitly recognize that price changes can regulate water demand but that such changes are, in turn, a measure of the scarcity of the resource. One-third of the developing world is expected to confront severe water shortages in this century due to increasing population size and changing climatic conditions. Subsequently, not only will the poor and the under-represented (e.g., non-human species surviving off ecosystems) have to struggle to find adequate water resources, they will have difficulty accessing safe drinking water free from the various forms of pollution previously mentioned.

Of course, similar to water access and pollution, there are mechanisms to mitigate water scarcity problems. In most parts of the world, temporal water scarcity is dealt with by building storage facilities, such as large dams and reservoirs that can be filled when natural flows are abundant and tapped when natural flows are scarce. When communities lack the space or political will to build new dams and reservoirs, some have developed conjunctive water management programs that capture surface water and store it in groundwater basins through recharge basins or injection wells, which are then pumped for use during times of drought. Similarly, communities commonly address spatial scarcity by building aqueducts and canals to move water from a region that is water-rich to a region that is water-poor. However, even when communities have the financial and institutional resources to build and maintain such infrastructure, a long-term drought or increasing water demand can render the infrastructure insufficient. The recent problems in the Southwest of the United States—a region that relies on Colorado River water to supply most of its growing population—is an example of this challenge. The main reservoirs in this area—Lake Mead and Lake Powell—dropped from being nearly full in 1999 to half-full in 2007, leaving states bickering about what actions they will take if long-term drought and growth continue. These states are now scrambling to find alternatives to make up for the likely losses in the basin they are so dependent upon. If structural solutions fail, migration can be the ultimate consequence of local water scarcity.

Some scholars and water managers argue that water markets or pricing schemes—especially in wealthy places like the U.S. Southwest—are an effective alternative for addressing water scarcity. In an economic system, the price of the commodity increases as scarcity increases, thereby regulating demand. However, pricing water is typically ineffective in modulating water supply and demand. First, there is political opposition to the creation of water markets because water is deemed a necessity. If water goes to the highest valued users, less wealthy individuals and species may be short-changed. Thus, even where markets exist, they are often regulated with pricing caps or only operate within limited sectors (e.g., between a small number of farmers). The legal and institutional frameworks that govern water allocation and rights can also create significant transaction costs that can hinder the effectiveness of even small-scale water markets. Such challenges indicate that although solutions to the scarcity are well-known, implementing those solutions is not easy when one considers the political, social and economic costs associated with alternative solutions.

## Clarifying the Global Dimensions of Water Scarcity

Given the difficulty of addressing water scarcity, access and pollution at even local scales, it is easy to understand why water crises are ubiquitous today. Yet, thus far, in characterizing the three water crises and their common solutions, we have not addressed any underlying global dimension of these three related problems. The global aspect of the climate issue is fairly obvious: Air pollution of individual countries translates into a pollution of the global commons that then impacts everyone in the future. However, we typically do not think of water as a global commons.

The possibility that North India may run out of groundwater in a decade leading to a collapse of agriculture in India is not viewed as a global problem. Likewise, the fact that the Yellow River no longer makes it to the sea, the fact that an aquifer in Long Island has been depleted and the three-hour daily walk for poor-quality drinking water in rural Ethiopia are all perceived and felt as local or regional problems. The discussion of global water crises refers to the vast number of people around the globe facing these problems. In essence, the global crisis is viewed as a collection of local crises—whether they are related to access, pollution or scarcity—for which there is a global policy, imperative. We rarely address the global elements of these individual problems. However, looking at the scarcity crisis more closely reveals a critical global issue.

### *Linking Local and Global Dimensions of Water Supply and Demand*

It is apparent to anyone who has studied or thought about the global hydrological cycle that local water availability is intimately tied to the global and regional climatic processes that control the disposition and movement of atmospheric and oceanic water. Thus, the climate is a direct bridge between local rainfall or water availability, and global processes. In an era of climate change awareness, this connection is now well-documented and disseminated through popular and lay media. However, the implications of this connection in terms of local and global water supply are not often obvious. At the global scale, it is possible that the hydrologic cycle may accelerate as climate changes, implying that rainfall patterns, and hence water availability, may change both in space and time. Additionally, climate phenomena—such as the El Nino Southern Oscillation—lead to concurrent and persistent droughts in large areas of the world. This implies that these areas are likely to experience water scarcity at the same time. With population growth, the ability of local storage infrastructure to buffer the population from the impacts of drought decreases. Further, noting that agriculture—and more specifically grain production—is a major water consumer, mega-droughts that span much of the globe limit the ability to address regional water scarcity through food imports from other regions. This emphasizes the dimensions of potential global water scarcity.

Many of our early population centres were built in areas with easy access to fresh water sources. Past civilizations that did not have continuous access to freshwater, or that existed in drought prone regions, sometimes perished (e.g., the Anasazi people of the U.S. Southwest). However, the development of local water storage and distribution infrastructure has since allowed societies to develop resilience to these climatic aberrations of supply. As a result, we still do not view water crises as stemming from a global hydrologic cycle that is akin to a global commons.

The ability to trade food is a primary factor that has allowed human populations to occupy certain geographical areas with much higher density than would be possible if all food had to be produced from locally available water. It is the virtual import/export of water through food that effectively connects the local dimensions of water scarcity to a global dimension. It is estimated that 30 percent of all water in global food today comes from a country other than the one in which the food is consumed. This fraction is anticipated to grow, meaning that global market forces will play a role in both the supply and demand for local water resources. As globalization makes food trade an implicit mechanism for reducing the impacts of local food and water scarcity, the vulnerability to water scarcity will eventually extend to a global scale. One consequence of this extension could be global agricultural price shocks. Another consequence could be new or increased competition for water at the local level among differing sectors—municipal, industrial and agricultural-competition that can play out in political debates, court disputes and/or conflicts.

If climate change and the associated human migration projections pan out as indicated by current modelling efforts, water surplus, low population density and low-intensity land use areas—such as Canada and Siberia—may emerge as population centres and the agricultural production centres of the 21st century. There is some recognition of the possibility of this trend. Yet, to our knowledge, no formal analyses exist on the effects that such changes will have on resource use—water, food, land or energy—and ultimately what the socioeconomic implications of such changing patterns in resource use entail.

### The Role of Agriculture

Agriculture is the dominant water user on the planet, accounting for 70 percent of global water use, on average, and greater than 90 percent in and or semiarid regions. Agricultural water use efficiency is typically very low. Although efficiency rates vary by crop type, for many crops only 10 to 20 percent of the water supplied in either irrigated (not drip) or rain-fed agriculture is transpired by the plant. The rest is lost either by direct evaporation from the soil or in the water distribution network. Furthermore, most of the agriculture on the planet is rain-fed, which has higher evaporation rates and lower crop yields than irrigated agriculture. Thus, given that agriculture is the dominant water use and dramatic reductions in agricultural water use are technically possible without an impact on food production, it is an obvious target for meeting the challenge of water scarcity.

If an order of magnitude reduction in agricultural water use could be achieved, there would be no global water scarcity for the foreseeable future, at least not as a constraint on global carrying capacity for humans and other life. This has led to slogans like "more crop per drop" and work towards identifying technologies such as drip irrigation or genetic modification of crops so they will consume less water or can be grown in salty water. Efficient irrigation timing—through weather monitoring, appropriate use of nutrients and the use of weather or climate forecasts—combined with appropriate cultivars, can also reduce water usage.

However, practical progress towards significantly reducing agricultural water use has been made in very few places, like Israel, where agriculture has shifted to drip or recycled waste water use and to high cash value crops from subsistence agriculture or the production of cereals. For instance, drip irrigation accounts for only about 1 percent of all global irrigation, even though it can substantially reduce water use relative to flood irrigation, which is the most common practice. Access to technology is only a small part of the solution. The economics, politics and sociology of agriculture—as well as education and cultural adaptation—play a significant role in limiting change. Globally, food and agricultural product prices were steady or declined in real dollars over the last fifty years, and are only now starting to increase.

Gains in productivity ushered in by technology—the Green Revolution—were responsible in part. Attempts to protect the rural sector—through water, energy and fertilizer subsidies—and support prices for agricultural products play a large and as yet, incompletely understood role in impacting local and global investment in reducing agricultural water use. For instance, an American cotton farmer who has access to advanced production technologies can achieve much higher yields than a cotton farmer in a developing country. Additionally, the American farmer, when given free or highly subsidized water, can then sell that cotton at a low price. Since the United States is a relatively large producer, it influences the global market price, thereby forcing other countries to politically provide similar support mechanisms to their farmers. This leads to profligate water use in both locations. The recent price increases reflect progressive limits on land and water productivity, the increasing population and the diversion of agricultural to non-food products like biofuels.

Subsidies have a role in development and could be redirected toward incentives for water conservation. This transition requires some degree of international policy dialogue and concurrence to address the inherent collective-action problem of responding to signals of water scarcity. However, in practice, internal politics decrees a maintenance and proliferation of the status quo. Additional supports are added to protect the sector from losses due to a lack of water available either due to upstream abstractions or due to a climatic exigency. For instance, farmers in the Punjab in India are lobbying for state funds to dig wells deeper, since the regional groundwater table has declined in places by about 1 meter per year, to a depth of about 400 feet. Added to that, energy for pumping is provided to groundwater users for free or at a highly subsidized rate. The result is that groundwater—which constitutes a fossil reserve—is being mined in many regions throughout India, and worldwide, where farmers have access to cheap energy and the financial resources needed to pay for increasingly deeper wells.

### *Towards Agricultural Water use Efficiency*

In countries such as India and China where population densities are high, especially in rural areas, agricultural water use already poses a stress on urban consumption. Major urban areas do not have the ability to provide drinking water on a sustained basis, even though they constitute a higher value use compared to agriculture and are able to pay for such use. For example, many major Indian cities face severe water shortages, often limiting public access to water from a few hours per week to a few hours per day. In theory, substantial volumes of water could be transferred from rural agricultural users to these urban sectors if greater efficiencies in agricultural use can be achieved. Much of the problem with such solutions in India is political. Providing highly subsidized rural water is a political norm that is difficult to challenge, even though the subsidies preferentially benefit a limited number of rich farmers and not the masses for whom these measures are intended. Improvements in the rural economy that facilitate agricultural water use efficiency are key for benefiting urban users and ecosystems as well.

One approach to achieving improvement in water use efficiency is to create a situation that leads to more revenue per drop. Increased financial resilience could then justify individual or group investment in a technology that facilitates a more efficient production of goods, while still assuring a high reliability of the water supply. Such resilience can further diminish the internal political pressure to support agricultural subsidies, driven in part by the need to protect farmers from financial loss. For example, higher revenue crop production has become available to some small farmers in India and China through the introduction of contract farming by national and multinational corporations.

Under contract farming, farmers are provided with all inputs and technological training to grow high-cash value crops that are exported or processed into goods and paid at the time of harvesting. The approach removes credit risk, market risk and technology risk from the farmer and provides incentives for higher yields from the same inputs. However, it requires a much higher technological and information base to execute through all stages of the product supply chain. This may lead to a transition toward higher efficiency production and a reevaluation of the time horizon over which the future value of the fossil groundwater resource or investments in improved surface water irrigation infrastructure are assessed. The corporation may be better placed to buffer the financial risks from which the farmer has been freed, and innovations on managing and sharing the climate risk through insurance and other channels are emerging. Finally, the corporation can look at the global marketplace for agricultural commodities and optimize what should be planted—where, when and how—considering market, labour and water constraints at local, regional or global levels.

## Global Awareness and Global Solutions

In summary, there is a water crisis in many regions of the world, and the problems will progressively become global. There are three crises, but the preeminent crisis is one of water scarcity One of the dimensions of global water crises is already obvious to many; the problems of water access, water quality and scarcity are felt nearly everywhere, most prominently with the world's poor and under-represented. When we focus on the issue of scarcity—which can drive the other two crises—it is easier to identify the global nature of these crises as well as the connection between global and local forces. One of the key global elements relating to scarcity is the role of climate change. While we are becoming more attuned to the impacts of climate change on local water supplies, to a large extent we have mitigated this global influence by developing infrastructure and other forms of adaptive water management tools. The biggest global force that currently drives water scarcity is agricultural water use. The global market for food contributes to the flows of water resources—at least virtual water—around the globe, intersecting with local economic forces that support inefficient water use within this sector. We argue that the solutions to the scarcity challenge—and ultimately to part of the access and pollution challenges—require both local and global action at the policy and economic level to improve agricultural water use.

The few solutions we have discussed in this chapter require significant policy changes at the local, national and international level. To remove the masks on water scarcity signals, nations have to find ways to either change or overcome the powerful interest groups in the agricultural sector. The incentives for any one country to shift policies are quite low without substantial international efforts to ensure that other major agricultural nations will also participate. If new agricultural supply-chain practices are part of the solution basket, industries need to be aware of these opportunities and national and local governments will need to find ways to incentivize and protect access rights for those farmers and industries willing to undertake new business practices. The precursor to such types of changes is, arguably, heightened awareness of the nature of the problem. Such awareness can occur if a professionalized forum is available to provide informed debate and opportunities for learning. Such an analytical forum is largely a public goods problem, and one that requires collective action at the international level.

The benefit of international efforts to collectively investigate global water problems is evident from the impact of the International Panel on Climate Change (IPCC). One might ask, "Why is it that climate change has become a much greater scientific and social concern than water scarcity, access to safe drinking water, water pollution or resource depletion?" Arguably, this is because the climate change research community has come to agreement on the nature of the problem and on a metric with which to measure it. Global temperature increase over the last century is now well-documented, and in the last two decades a lot of scientific and political effort has been expended to connect this to greenhouse gas emissions, deforestation and other human-induced changes. Model-based scenarios and scientific intuition are used to project dire consequences for the future of the planet and its societies if greenhouse gas and temperature trends continue. A synergy between scientific inquiry and political action drives the climate change agenda further on both fronts as a global issue. Many significant uncertainties remain with climate change projections, even though the causative factors are agreed upon. A large number of these uncertainties pertain to the details of the hydrologic cycle—how water availability will change and how the distribution of water through the atmosphere and through vegetation will modify climate. So far, we have only seen the beginnings of the conclusive impacts of climate change, and the action agenda is to find solutions before this crisis overwhelms us and causes irreversible damage. Our best guess is that these impacts may occur over decades.

Addressing climate change requires policy action, but also a search for technical solutions to quickly substitute fossil fuels, or effective carbon capture and sequestration. Energy conservation and use-efficiency improvements are important but will not be sufficient to meet the challenge. These observations emerge after nearly two decades of intense global debate—one that has been fueled and supported by information generated through extensive international research efforts—as well as the synthesis and scientific consensus coming out of the Intergovernmental Panel on Climate Change. The debate continues, but at a relatively high level of maturity, and rapid technological evolution to address the anticipated problems is in the offing.

By contrast, the dimensions of the current and emerging water security challenge are not as well understood. Even when attention has been drawn toward water problems, it has been fragmented across the three types of crises we face today and targeted towards the local dimensions of the problems, which incidentally, are similar to the symptoms of global energy and climate change issues. This has prevented a focus on the preeminent scarcity crisis, which requires much more information generation and understanding. It is in many ways a more complex problem than $C[O_2]$-induced climate change, given that its global and local causes are more closely identified with social factors, local access and supply chains, and with many layers of common pool resource problems. Developing a single technical solution that will address the problem is at first difficult to recognize. However, the agricultural use-efficiency issue emerges quickly as a dominant issue. Given that at least some technologies are readily available for use reductions, the solution in this case may very well be to first achieve conservation and efficiency improvements using these technologies. Innovations in corporate farming offer some promise to improve rural livelihoods, while providing access to global markets for agricultural products and facilitating the reduction of climate, market, credit and labour risks through efficient global pooling of financial resources. New technologies to reclaim contaminated or saline water may also be an important part of the equation. This could involve using solar or wind technology as energy sources or coupling them to efficient greenhouse-based agriculture. Greenhouse-based agriculture—which has been widely used in Israel—can produce high crop yields while recapturing much of the water transpired by the plants and evaporated from the soil. The reuse of water can lead to substantial reductions in overall water use from the agricultural sector. Current pessimism as to whether we will have sufficient water to support life on Earth could be transformed into optimism, if we could formulate a path working toward these types of innovations.

The formation of a global roundtable as a focal point for analysing the causes and anticipated future conditions of water scarcity could help achieve what the Intergovernmental Panel on Climate Change has done for climate change. An interesting difference in this regard is that the political dimensions and social importance of the water issues are much better recognized at the outset than they were for the climate change problem at the initiation of the IPCC. On the other hand, the causal structure and scientific bases for climate change and its impacts, as well as the supporting databases, were perhaps much clearer for climate change than for the water scarcity problem. Consequently, a global roundtable would need to focus, from the beginning, on the collection and analysis of the wide range of information sources that are relevant to understanding and predicting the causal structure of local and global water scarcity. It is through such a process that both the scientific and policy communities can begin to understand and respond effectively to the global drivers of water crises, fostering tools and choices that could help leverage the diverse and important local solutions that are already available.

# Chapter 6

## Perspective in Water Management of Megacities

Megacities—characterized as metropolitan areas of 10 million or more inhabitants—have become an important focus in terms of water provisions, sanitation services and the related impact of urban development on natural resources. While basic needs of residents of smaller cities are very similar, the emphasis placed on megacities lies in the fact that these massive urban conglomerates have grown to almost unmanageable dimensions. This, in turn, has made water provision and sanitation services to the entire population a bleak, if not unachievable, task. In most megacities of the developing world, clean water is both scarce and expensive to produce.

Large sectors of the population in such densely populated urban areas do not have access to potable water, and even larger sectors of the population do not have access to sanitation. Non-revenue water—water that is produced and enters the distribution system but that is never billed to consumers because it is lost due to leakages or illegal connections—is up to 30 to 40 percent. Infrastructure is either scarce, becoming complex or deteriorating. In addition, water supplies are largely underpriced and necessary investments are estimated to reach billions of dollars. However, as complex as these matters are, the real challenges in terms of water provision point in another direction. The main problem in urban concentrations, which is further heightened in megacities, is the lack of appropriate management, adequate institutions and sustainable planning to address these challenges beyond short-term approaches.

To provide an overview, this study first analyzes the nature of megacities on a global scale, addressing the delicate relationship between economic development and urban resource management in the face of environmental sustainability. This topic then unfolds through an analysis of the Metropolitan Area of Mexico City (ZMCM), a megacity accounting for approximately 20 to 22 million people and recognized for its diminishing supply of natural resources, among them, freshwater. Emphasis is placed on declining groundwater supply, complexities with infrastructure and concerns regarding sanitation. While the central focus of this argument is the ZMCM water supply; a comparative analysis points out the similarities and differences among megacities across the globe, such as Dhaka, Sao Paulo and Bangkok. Ultimately; this chapter attempts to demonstrate that the problem of water access in megacities has less to do with water scarcity and more to do with water mismanagement.

## Megacities as a Growing Phenomenon

As urban growth continues to increase worldwide at escalating rates, megacities remain a growing phenomenon. According to the Organisation for Economic Cooperation and Development (OECD), as globalization progresses, urban regions—in particular, cities with populations exceeding 10 million people—are playing increasingly important roles in the global economy in terms of competitiveness, thus becoming major centres for both local and national policies. Yet, most megacities have focused on economic growth with almost no consideration for environmental sustainability.

As a result, these extensive metropolitan areas have created alarming levels of water and air pollution, solid waste and deterioration of their own natural resources. These problems have also reached surrounding regions affected by urban spraw. The most serious environmental problems are associated with the disorganized expansion of legal and illegal settlements, over-exploitation and degradation of surface and groundwater sources, air pollution, solid waste and the indiscriminate destruction of natural resources. This has created what seems to be an endless vicious cycle in which the deterioration of the environment has resulted in social erosion and vice versa. Mexico City and its surrounding areas have not been an exception.

While Mexico City has achieved high economic growth, its development is still deeply polarized and asymmetric. Economic and social development has flourished in some parts of the megacity resulting in prosperity, but inequalities have become increasingly acute in other parts of the region. In order to maintain and improve the living standards of a growing population, major investments are needed to protect basic resources. Failing to address environmental degradation—such as shrinking freshwater sources—is counterproductive to economic development in the long-run. This holds true because while most megacities focus on increasing foreign competition and employment at the cost of their natural environment, foreign investors tend to be most interested in megacities that not only have the most appropriate institutions, infrastructure and human resources, but also liveable environments and natural resources that will sustain economic activities.

Although all megacities face similar challenges in terms of access to natural resources—like clean water and ensuring adequate control of air, land and water pollution—these urban conglomerations differ in terms of growth rates, types of population, infrastructural needs and environmental conditions. In the developing world, metropolitan areas tend to have high population growth, consistent needs for infrastructural development and increasing deterioration of their environment. In contrast, in the developed world, megacities tend to have slower urban growth, infrastructures that already need to be replaced and more regard for the protection of their environment. In both developing and developed megacities, migration of formal and informal settlers plays an important role in terms of demand and provision of services. In Mexico City and its surrounding areas, illegal settlements are a critical problem. In 2003, there were 804 irregular settlements in Mexico City with approximately 60,000 families. In Mexico State—the most populated of the country's thirty-two states—illegal settlers are creating mushrooming cities that lack basic services. Much of this population is established not only in valleys and hills, but also in areas prone to flooding, which can result in flood hazards that can destroy entire communities. Legal and illegal settlements in flood-prone areas and their social and economic related costs have been heavily felt in several other megacities in addition to ZMCM, such as Sao Paulo and Dhaka. In most cases, it is a problem of which control and solutions go beyond the water sector and which is vested in land-use planning institutions.

Due to their differences in terms of population, location, climate, infrastructure and environmental concerns, each megacity needs to plan and implement its own tailored solutions at its own pace according to local economic, social, environmental and cultural conditions. Yet, a common denominator uniting megacities—mainly in the developing world—is that rates of urbanization have surpassed the capacities of both local and national institutions to provide the population with the necessary services. According to the UN Population Division, the number of megacities has increased from sixteen to eighteen from 2000 to 2005 alone, and this number is predicted to reach twenty-seven by 2025. It is noteworthy that the majority of megacities at the global level are located in the developing world.

In spite of urban growth, the percentage of urban populations living in megacities worldwide has remained about 8 percent from 2000 to 2005. On the contrary, it is in urban conglomerates of less than 500,000 people where the percentage of urban population at the global level is still the highest (52 percent) by cities with 1 to 5 million inhabitants (22 percent). This means that, despite the rate of increased urbanization throughout the last decades, the highest percentages of the world population do not live in megacities; they are more likely to live in small or medium-sized cities. While the growth of small and medium sized-cities is projected to escalate and, with it, the challenges of providing adequate water services, it is the megacities that have turned to almost unmanageable dimensions and require a reexamination of water management.

Furthermore, megacities are not restricted to their geographical boundaries but extend to the regions surrounding them. The escalating demands of megacities in terms of clean water, among other basic necessities, have resulted in negative impacts on the surrounding states and regions, which have been forced to supply megacities with additional commodities. Since megacities continue to increase in size, demand requires the implementation of development models based on social equity and environmental considerations, not only at the local level, but also at the regional level.

Despite these bleak realities, alternatives exist to improve the governance of megacities in order to provide citizens with better access to clean water and other services, as well as improved environmental conditions. OECD, for example, suggests that governance-related issues should be addressed at the metropolitan area level in order to maximize the potential of the urban region. Some of the actions mentioned to improve metropolitan governance include encouraging local governments to work in partnership with business communities, introducing greater transparency and accountability in decisionmaking, developing better frameworks for long-term strategic planning, and improving public access to information and public services, all of which could lead to positive results. The limitations that OECD identifies for achieving better governance at the metropolitan level, however, include a fragmentation of the administrative jurisdictions, a strain on the financial and fiscal ability of local authorities as well as the lack of decisionmaking processes that are both transparent and accountable. There is also a lack of ability to coordinate and effectively communicate between cities that form part of metropolitan areas.

## Complexities of Water Resource Management in Megacities

From the late 1990s to the present, water management trends have included, at least in theory, the implementation of concepts such as sustainable development, river basin management and integrated water resources management approaches. An emphasis has been placed on multi-sector approaches, as well as higher considerations for social and environmental issues, with increasing importance given to governance-related issues.

In the water sector, attempting integrated approaches has made water management goals even more complicated to implement than before, particularly in the context of megacities due to their sheer size. With the added expectations on implementation of water governance issues, even more complex processes have been introduced—such as transparency and accountability, or multilevel participation and multilevel decisionmaking—even when the most basic and pressing needs in terms of water supply and sanitation have not yet been solved.

Megacities require large investments to cover the needs of all sectors of the population in terms of water supply, drainage and sanitation. However, in addition to the enormous investments, a main constraint for the efficient provision of water supply is that governments have multiple, fragmented institutions with responsibilities that are often not clearly defined or overlap with each other. This makes any long-term decisionmaking almost impossible, leaving strategic planning in the utopic realm, resulting in escalating degradation of the urban environment.

Although considered basic needs, water supply, drainage and sanitation are not always an investment priority in megacities. This is because lack of water, both in terms of quantity and quality, does not affect the entire population to the same degree. In general, wealthier populations have access to tap water in their houses, as well as functioning drainage and sewage systems, while less affluent sectors of the population depend on standpipes or purchase their water from vendors. Many poor regions are either connected to a public sewer or septic tanks, or have to dispose their wastewaters in surrounding rivers, lakes or even on their own property. Lack of tap water and drainage in megacities, therefore, is a problem faced by less affluent sectors and is therefore their problem alone. Sanitation is an increasingly serious issue which only a select few communities have to face directly but which affects the health of the entire region.

As previously stated, water management also faces the complexity of flood-prone zones. For example, the number of humans exposed to floods has tripled during the last two decades to almost 2 billion at present. The main reason behind this problem is the concentration of hundreds of millions of people in mushrooming cities in both deltas and floodplains. On the other hand, megacities sprouting in arid and semi-arid regions face water scarcity on a daily basis, not only because of mismanagement but also because of an actual water scarcity problem to begin with. Regarding water for irrigation, megacities devour enormous amounts of provisions which have to be transported from the countryside, often at great distances.

### *Megacities and the Escalating Case of Water-related Challenges*

The debate on water resource management in the international arena most often focuses on integrated water resources management, river basin management, good governance and public participation. In spite of this rhetoric, priority in megacities is the provision of clean water even at increasing social, economic and environmental costs. Investments for water supply and sanitation in the developing world, including megacities, has escalated considerably. In the case of Mexico, water supply and sanitation investments were estimated to be $1.7 billion per year in 2002. In 2005, China invested a total of $9.7 billion on flood control, infrastructure, soil, water and ecological projects and hydropower development. At this point, many scholars and water activists have pointed out the failures of achieving the development goals at the international level; cities provide water that is not always potable, and collection of sewage has been conveniently considered as sanitation even when it is not treated.

Varis et al. summarize the situation and future prospects of seven rapidly expanding megacities in various parts of the developing world: Jakarta, Johannesburg, Silo Paulo, Mexico City, Riyadh, Istanbul and Singapore. In their analysis, the authors conclude that megacities are dramatic cases of urbanization and water-related challenges. A main constraint is that even though the provision of water for various sectors is quite important, water is not the only aspect of infrastructure development needed in megacities. Transportation, energy and housing should also be developed.

The water footprint of a megacity goes far beyond its city limits. Megacities import food from the international food trade, the bulk of which has been produced using water irrigation practices. Megacities also import massive amounts of other resources such as energy, metals and fiber products, which can interrupt hydrologic systems, both qualitatively and quantitatively. In order to be sustainable, megacities need flourishing economies, strong social and environmental policies, proper governance systems and adequate mechanisms for public participation. All of these aspects are intertwined with the various roles that water plays in megacity development, and the water sector is often an important contributor.

Finally, economic growth, as well as social and environmental factors, should be considered simultaneously Competitive megacities are better able to attract resources to invest in infrastructure, thereby providing quality services for the population and, in turn, better economic and social opportunities for large sectors of their population. The challenges to achieving this are immense. However, initiating the necessary processes and frameworks for financing, managing, maintaining and providing efficient services, as well as finding long-term solutions that combine social, economic and environmental view points, is necessary. Otherwise, freshwater access and sanitation become unmanageable.

## Water Governance in the Mexico City Metropolitan Area

Applying this overview of megacities to a case study, this section offers examples of the complexity of clean water services faced by the ZMCM. It is important to stress that, more often than not, inefficient, costly and unreliable water supply services in Mexico City are the result of inappropriate management practices, not merely the result of an actual scarcity in water supply. To reiterate the size of this megacity, the ZMCM has an approximate population of 20 to 22 million people. It generates approximately 35 percent of the national GDP in Mexico, which makes it strategically quite important from both an economic and political standpoint for the country. High rates of migration, in addition to rapid urban development, have made Mexico City not only one of the largest cities in the world, but also a city with one of the highest population growth at the global level. The phenomenal growth rate witnessed between 1950 and 2005—from 2.9 million people to more than 20 to 22 million at present—has made it almost impossible for institutions to respond to the rapidly increasing needs of the population in terms of water supply, drainage and sanitation.

The rapid growth of the ZMCM has been the result of policies that have favoured social services, long-term major investments in all sectors (energy, water, transportation, telecommunications, etc.) as well as heavy subsidies and industrial production. Even though government planning strategies have been focusing on decentralization efforts for decades, tax subsidies, as well as numerous social, political, educational and cultural factors, have made the city more attractive than other regions within Mexico.

The multiple levels of public sector institutions that are involved in the solution of water-related problems make decisionmaking a very complex process not only in the ZMCM, but in all megacities. In Dhaka, a city with approximately 12 million inhabitants, organizations responsible for water supply include government institutions at the federal, city and municipality levels, the private sector and even donors. The World Bank, for example, was the main donor of the Fourth Dhaka Water Supply Project and insisted that changes in governance and management be implemented at both the Dhaka Water Supply and Sewerage Authority (DWASA) before the credit was approved. In the case of the Silo Paulo Metropolitan Area, decisionmaking in water-related issues for their 18 million people involves federal, state and municipal levels of the city of Sao Paulo, as well as the additional thirty-nine municipalities that form part of the megacity.

The ZMCM includes two different political entities and their several levels of government. Decisionmaking depends on decisions based in the sixteen boroughs of Mexico City, thirty-seven municipalities in the Mexico State—each municipality consists of an independently elected representative—and the private sector companies which are also involved in service provision. Another important issue is that from 1997 onward, the citizens of Mexico City were able to elect their own government. Prior to this, the mayor of Mexico City was appointed directly by the president, who, from 1929 until 2000, belonged to the same political party The government of Mexico City, therefore, has often been considered an extension of the federal government, with criteria and priorities mirroring that of the federal government. This means that, for better or for worse, there was agreement on the policies that would be followed in the metropolitan area. Currently, the different political parties represented in the governments (federal, state, boroughs in the case of Mexico City and municipalities in Mexico State) have diverse interests, priorities and criteria. This diversity, while unquestionably positive in terms of democracy, has resulted in a more serious lack of coordination and agreement in terms of economic, social and environmental policies. The reality is that political criteria prevail even when this is not always positive for the megacity.

## Groundwater Supply in Mexico City and Beyond

The added complexity of the ZMCM for its water management, in addition to its large population and fragmented government, is that it is located in a naturally closed basin, making it vulnerable to floods. In addition, its own rivers have become wastewater canals. Furthermore, increasing subsidence has resulted in extensive damages to the water supply and sewerage infrastructure, as well as degradation of the quality of the groundwater on which its own survival depends.

The water supply in ZMCM will provide for 20 to 22 million people; the economic sector depends primarily on local groundwater sources (70 percent) and on inter-basin transfers from outside basins. Infrastructure for distribution of water within the city has become very large and complex. It is comprised of sixteen dams having a total storage capacity of more than 2,800 cubic kilometres, a primary network with more than 1,000 kilometres and a secondary network with more than 12,000 kilometres.

In order to supply the necessary water for the needs of the social and economic sectors of the megacity, the rates of extraction from the aquifer have been much higher than the recharge rate for decades. At present, approximately 45 to 54 cubic meters per second is abstracted per year, but the natural recharge rate is only about half of this figure. This has resulted in very serious overexploitation, with the groundwater table decreasing by around one meter each year. It is estimated that approximately 670 legal wells exist for groundwater extraction.

Likewise, the megacity of Dhaka, Bangladesh, depends heavily on groundwater. The water supply system for the city that was constructed more than one hundred years ago was based on surface water. However, the cheaper and easier option was the availability of groundwater, which resulted in a supply system based on groundwater. At present, with surface waters heavily polluted, institutions face the challenge of providing services to a population increasing at approximately 6 percent per year, while simultaneously trying to reduce dependence on groundwater. This is a system that is clearly unsustainable.

Available data indicate that the groundwater table is falling at an alarming rate of more than 2 meters per year. Groundwater is being mined in significant amounts. Consequently, the discharge of the tube wells are being reduced at a much faster rate resulting in higher cost of pumping and drying up of the bore holes. The life of bore holes has been reduced from twenty years in the late 1960s and early 1970s to less then five years now. Investment in new deep wells is quickly becoming economically unattractive.

Deficit between supply and demand is expected to widen again from 2004 but at a much faster rate due to the inability of the DWASA to secure necessary funding for the construction of two more surface water treatment plants. At the same time, construction of new and replacement old deep tube wells will not produce designed discharge as the groundwater reservoir of both the primary and secondary aquifers have nearly been exhausted. Though a new initiative of exploiting deep aquifers (more than 300 meters deep) have yielded some promising results in a certain area of the city, any large scale development of the deep aquifer will have to be carefully evaluated for its technical, economic and environmental viability.

According to official figures the water extracted from the aquifer to supply water to the population of Mexico City from 1990 to 2007 has reached almost 17,000 cubic meters per second. This figure is significantly higher that the groundwater recharge rate, illustrating how this trend in groundwater extraction is unsustainable.

The problems related to the extraction of groundwater in Mexico City are not new. Several issues relating to the city's survival were identified decades ago. Since soil is composed of clay, the more water that is abstracted from the aquifer, the more susceptible the soil is to compaction. Subsidence of several parts of the city has resulted in extensive and costly damages to both surface and below ground infrastructure, including water supply and drainage.

The ZMCM is not the only megacity which suffers from severe subsidence. Bangkok is also affected by subsidence in certain areas of the city. In the case of Bangkok, subsidence rates are estimated to be 1.5 to 2.2 centimeters per year as of 2001, a much lower figure compared to that of one decade ago which was 5 to 10 centimeters per year. In comparison, the central area of the ZMCM has subsided approximately 10 meters during the 1900s, varying from 10 centimeters per year in the central area to 20 to 25 centimeters in the area of the International Airport of Mexico City.

Both megacities have established mitigation measures in the attempt to reduce the abstraction rates of their aquifers. These have included policies, economic disincentives and, in the case of ZMCM, protection of conservation or rural areas for natural recharge purposes. However, conservation areas are under increasing threat because of steady growth in both urban and rural areas. Regarding rural growth, 35 of the 44 rural settlements that still exist in Mexico City and which include 400,000 people, are also located in conservation areas.

## Access to Tap Water in Megacities

In terms of access to water, in the developed megacities, such as in London and New York, all residents normally have access to tap water in their houses. In the developing world, a common problem in most megacities is that millions of their inhabitants do not have access to clean water in their houses or nearby. These populations depend on either standpipes, vendors or provision of water from tankers provided by the public sector. In these cases whether or not residents are consuming good quality water is uncertain.

In ZMCM, as in the Silo Paulo Metropolitan Area, approximately 90 percent of the population has access to water, either through a water connection directly to the house or from common faucets in the neighbourhood. In the case of ZMCM, most of the water sources are located to its west, north and south. Thus, the water supply is irregular and unreliable for the population living in the eastern part of the city, the subsection most affected by water shortages. An alarming figure is that in Jakarta, only approximately 41 percent of the population had access to tap water in 1996, a figure which increased to 56 percent in 2003 despite the changes in the institutional arrangements and the involvement of the private sector.

Non-revenue waters are one of the main challenges facing megacities both in developed and developing countries. The Office of Water Services (OFWAT)—a water regulatory agency of England and Wales—provided Thames Water (a private sector company) a target to reduce water losses to 805 million liters per day for 2006-2007. In Mexico City and its surrounding areas, non-revenue water is approximately 30 percent. In the case of Tokyo, on the other hand, non-revenue water accounts for only 9 percent. Lack of investment tends to be the most frequently mentioned reason for high, non-revenue water. However, lack of managerial skills and manpower, as well as inefficiency and poor planning, are also considered major causes for the low performance of utilities providing the water supply.

Unreliability of water quantity results in extra expenditures for certain sectors of the society in ZMCM. In fact, it is estimated that the population without access to tap water pays 6 to 25 percent of their daily salaries in purchasing 200-liter containers of water. Nevertheless, mistrust in the quality of Mexico City's tap water also has negative impacts in terms of extra expenditures for all sectors of the society irrespective of their economic status. Drinking water for most of the people in the ZMCM, even when they have access to tap water, comes mostly from 20 to 30 liter containers of drinking water that are sold commercially. A main problem, however, is that companies that bottle water in the ZMCM are not monitored and thus water quality of the containers is not always in accordance with the specific norms and regulations.

The perception that the quality of tap water is not suitable for drinking purposes has an enormous impact on the economy of the country. Mexico is the second largest consumer per capita of bottled water in the world, with consumption increasing from 11.6 billion liters in 1999 to 17.7 billion liters in 2004. At present, per capita bottled water consumption in Mexico is nearly twice that of the United States, even though its per capita GDP is about one sixth that of the United States.

The Third World Centre for Water Management has carried out analyses on water quality in Mexico. According to this study, in 2005 alone, the expenses due to waterborne diseases in the country were estimated to be $260 million, with the highest mortality in population between zero and four years of age and above sixtyfive years of age. Nevertheless, despite the economic importance of the impacts that water-borne diseases have had on the country for decades, these findings have not been documented thoroughly and addressed properly by health and water-related institutions. In summary, the problems related to the drinking water supply in megacities include multidimensional, multi-sectoral and multi-regional issues, which depend on the interests and capabilities of more than one institution and which require multi-stakeholder coordination.

## Wastewater Collection

In addition to water supply, the other main challenge for megacities in terms of water is collection of wastewaters and their proper treatment and disposal. It is said that 2.6 billion people lacked improved sanitation facilities in 2002, which is a number too big to be ignored by governments and by the international community. Not even the United Nations considered sanitation important enough to be included in the Millennium Development Goals. It was not until the World Summit on Sustainable Development organized in Johannesburg in 2002 that sanitation goals were considered as a part of a global effort to improve the quality of life of the populations as well as quality of the environment.

In the ZMCM, about 50 cubic meters per second of wastewaters are disposed without treatment. Almost all of this water is used for irrigation purposes, resulting in very significant health and environment-related problems and concerns. In 1996, the Inter-American Development Bank approved a $1,035 million project for the Mexico Valley Sanitation Project.

Unfortunately, this much-needed project did not proceed, mainly due to economic and political reasons. In 2004, water institutions from Mexico City, Mexico State and institutions at the federal level were working jointly with the Inter-American Development Bank (IDB) and the Japan Bank for International Cooperation (JBIC) to develop the terms and references related to the construction of four wastewater treatment plants. The total budget for this project was approximately $1 billion, of which IDB would contribute $365 million for the collectors system, the JBIC would provide $410 million for the wastewater treatment plants and the federal government, Mexico City and Mexico State governments would provide $260 million.

This enormous sanitation project was abandoned for years by the federal government, Mexico City and Mexico State authorities, being reconsidered only recently. Seven treatment plants, instead of the seven mentioned initially, are now expected to be constructed with a total capacity of 40 cubic meters per second. The approximate investment will be $2,345 million. Only time will tell if the project will be completed.

## Energy Consumption and its Relation to Water

Concerning energy, megacities require massive quantities of power. Megacity residents consume about five to ten times the energy of the national average. In addition, all large-scale electricity generation requires an enormous amount of water, either as hydropower or for cooling purposes, as do bioenergy sources, making water an important prerequisite to satisfy the energy, requirements of megacities.

For the ZMCM, the daily consumption of fuel in 2004 was estimated to be 306,000 barrels equivalent of gasoline or petrol, or 48.6 million liters per day. In 1990, the estimated figures were 37 million liters per day, that is, there was an increase of 31 percent in fourteen years. Drinking water to Mexico City has to be pumped to a height of more than 1,000 meters. Equally, because of the increasing soil subsidence, wastewater also has to be pumped up to discharge from the city. The above has made water supply and sewerage very energy-intensive and expensive. In 2000, for example, energy used in the ZMCM on pumping clean water, treating water for drinking purposes and collecting wastewater and rainwater was 2.436 Peta Joules. As it can be observed, the necessary energy needed to pump clean water has cost the country almost $600 million per year between 2000 and 2005.

### *Strategies for Drinking Water and Sanitation: Lagging Behind the Needs*

Strategies to increase water supply and sewerage services have focused mainly on infrastructure development throughout the years. The different strategies (e.g., the 1995 Master Plans for Drinking Water and for Sewerage for Mexico City) have outlined the necessary infrastructure to improve supply, storage and transportation of drinking water in Mexico City, as well as the storage, transportation and disposal of wastewater and stormwater.

For several years now, the program on drinking water for the ZMCM has considered constructing new infrastructure to transfer more water to the metropolitan area from even more distant and expensive sources. These alternatives have considerable economic, social and environmental costs.

In the case of the transfer of water from Cutzamala River, the construction of its fourth stage would have an initial investment of $502 million to increase the volume of water only by 5 cubic meters per second, from 19 at present to 24 cubic meters per second. If the water transfer from the Amacuzac River was implemented, in addition to the cost of the infrastructure, the annual electric power consumption for this system would be about 5 percent of the annual national electricity production, representing 16.5 million barrels of oil per year. In terms of investment, each cubic meter of water from the Cutzalama River is estimated to require an investment of $23 million. This figure would increase by a factor of four if the source of water were the Amacuzac River. This is a very clear example of the costly impacts that the disorganised growth of the megacity has imposed in the surrounding regions, creating acute economic, social and environmental problems.

Considering the above, it will certainly be more acceptable from the economical, social and environmental view points to implement demand management practices like reduction of unaccounted losses, water pricing and other water conservation practices, before embarking upon extremely expensive new water development projects, with high social and environmental costs.

It is estimated that the large volumes that are lost due to leakages in the distribution systems in the ZMCM—representing between 30 and 40 percent of the total volume of water that it is distributed—are second only to human consumption. It would thus be preferable to focus first on repairing the distribution networks before embarking on such expensive projects, especially if water would still not reach the consumers due to the losses.

As Iracheta mentions, water in the metropolitan area is both scarce and expensive. In spite of this fact, water is not reused or treated after use but discharged raw, with the associated environmental and health-related problems, to another basin. Unfortunately, even then, long-term and rational planning, as well as coordinated policies for the development and management of the metropolitan area, including water resources, is still not the priority for the federal government or the government of Mexico City and Mexico State.

In 2003, the government of the Mexico State filed a lawsuit against the federal government and the government of Mexico City. This occurred as a result of the lack of response from the federal government and city government to the request of the government of the Mexico State to renegotiate the agreements to transfer water to Mexico City from one of its sources, the Lerma River. The lawsuit demanded the payment of $2.5 billion as compensation for the use of water from the Alto Lerma from 1970 up to that moment. It also demanded that the federal government become responsible for the operation of the infrastructure, which is operated by the government of Mexico City since the construction of the projects even though they are located in Mexico State.

Finally, it demanded that the federal government implement actions to recharge the aquifers. Even though Mexico City pays the federal government for the rights for the bulk water it receives from the Mexico State, the lawsuit from the Mexico State asked for compensation for the environmental damages that have been caused due to the overexploitation of the Lerma River. The decision of the Supreme Court was expected to set precedents for similar cases in the future. However, in October 2005, the newly elected governor of the Mexico State publicly declared that he would withdraw the lawsuit, since he preferred to work with the federal and Mexico City governments to find an amicable solution.

## Solutions on the Horizon: When Efficiency Overcomes Size

While the size of the geographical area does matter, so too does the planning and effectiveness of decisionmakers. There are two urban conglomerates in the developing and developed world, respectively, which have lately improved the provision of water supply and sanitation to their populations to levels higher than any city in the United States or Western Europe, irrespective of their size in terms of population. These are Phnom Penh, Cambodia and the country of Singapore.

In the case of Phnom Penh, in 1993, the non-revenue water was 70 percent, the tariffs (as is typical in all developing country cities) were lower than the cost of water produced and the water utility faced serious economic and human resources constraints. Their billing system was quite inaccurate and the actual volume of water billed represented approximately 28 percent of the production. From those actually billed, the collection ratio was only about 50 percent. Finally, the total annual income was only about 0.6 billion Rids, against an operating cost of 1.6 billion Riels.

By 2006, the Phnom Penh water utility developed an operational billing and collection system through which the customer database increased from about 20,000 in 1993 to almost 160,000 in 2006, and the collection ratio increased to 100 percent. The tariffs have been increased to more realistic values, and non-revenue water is now less than 10 percent (lower than London and similar to Tokyo). The difference in Phnom Penh was achieved as a result of leadership, political support, managerial skills and human resources development.

The water supply in Phnom Penh provides clean water now to 100 percent of the population, a striking difference from only ten years ago.

Singapore is an exceptional example of what planning and management, as well as long-term vision, can achieve in a period of time as short as forty years. Singapore, being a water-scarce country due to a limited amount of land where rainfall can be stored, has developed—and implemented—innovative solutions that have included both demand and supply management strategies. Demand management strategies have included water transfer from the state of Johor in Malaysia, management of their reservoirs, desalination and development of very high technology to treat used water and produce very high quality water, known as NEWater. Supply management strategies have included conservation practices, tariffs to promote water conservation and large public education campaigns. Overall planning and management strategies developed in Singapore have not been matched anywhere in the cities of the developed world, irrespective of their size.

It has been said that solutions in Singapore can be implemented because of the small size of the city-state. Nevertheless, supply and demand management strategies are, or should be, common practice everywhere, especially in megacities where the needs to properly manage the water resources available is more pressing due to the size of the population that has to be served. For water policies to be useful for the cities and their population, these management strategies have to be implemented. In only a few cases is there the will, the skills and the vision to implement them. Singapore is one of these few cases.

The supply of clean water and sanitation services in megacities should be viewed as an integral component of rapidly expanding urban areas. Clean water and sanitation services should not be considered in isolation, but instead in relation to issues such as their geography, climate, population growth, urbanization, migration, economic development and social expectations. While most of the cities in the developed world are growing slowly, cities in developing countries are growing in a disorganized and unplanned manner, resulting in heightened pressures to provide water supply and sanitation services to their ever-growing populations.

Provisions of services do not only include operational alternatives, such as the development of massive infrastructure, which is a necessary, but not always ideal, alternative. They also involve the development of a very complex series of issues that have to be adapted to each case, which include management aspects, governance concerns and decisionmaking adaptations—mostly shared between several actors—to suit the local conditions, develop a basket of economic instruments as alternative measures and encourage human resources skills to accelerate socioeconomic development.

An issue that is normally disregarded, but is of high importance, is that economic activities and naturally available water resources—rainfall, springs, lakes or rivers—do not always go hand in hand. In addition, most megacities have either exploited or polluted their surface and groundwater resources, making it necessary to transport water—many times hundreds of kilometres—at increasingly higher economic, social, environmental and political costs in order to cover their own needs.

Finally, megacities are faced with challenges that require immediate action. One of the most important challenges is governance. Governance-related issues are not just the domain of the government any more. The increasing participation of the private sector and societal actors—who are willing, and also demanding, to participate in solving problems that affect them—has made decisionmaking very complex. These complexities increase exponentially within the context of a megacity. Nonetheless, improving water governance in megacities should be of the utmost concern.

## Water Management in the Upper Little Colorado Watershed

Perhaps nowhere in the United States has climate and water resource interaction been more closely observed than in the U.S. Southwest, and with good reason. Overall, semiarid to arid conditions predominate. These conditions, together with wide variability in temperature and precipitation over time and space, challenge even the most earnest human efforts to sustain livelihoods and communities. Scientific studies and popular literature alike reflect recurrent episodes of overexploitation of natural resources due to inadequate recognition or acceptance of environmental limits such as water availability.

Accretion of local knowledge and experience provides a mechanism for anticipating and adapting to the vicissitudes of variability and change. However, factors such as reliance on faulty heuristics (Nicholls 1999), occurrence of events falling outside the range of local memory, policies that restrict decision options, and narrow focus on immediate economic returns can render such knowledge ineffective at best and destructive at worst. Science-based knowledge and predictive capabilities provide avenues for improving decision processes in the face of interacting environmental and societal stresses. To be accepted and used appropriately, such knowledge needs to be franked in ways that mesh well with local values and practices. Environmental histories that create a contextual account of embedded experience, values, and practices provide insights useful to this task.

Recent extended dry conditions in the U.S. Southwest have prompted efforts by researchers and forecasters to develop climate information products useful at local and regional scales for addressing drought impacts. This chapter provides an environmental history of climate and water resource management for the Upper Little Colorado watershed in northeastern Arizona, with a focus on the hundred years between the mid-nineteenth and mid-twentieth centuries. The study, funded through the NOAA-funded Climate Assessment for the Southwest (CLIMAS) project, provides insight into the experiences, values, knowledge, and practices that have produced conditions existing today. This type of contextual information, if integrated into development of plans, decision processes, and decision tools, holds promise for improving citizens' acceptance of and support for policies aimed at ensuring the sustainability of the watershed's closely coupled human and natural systems.

Based on established historiographic methods, this study aims specifically to uncover how communities and individuals in the watershed have developed and managed water resources in the context of climate variability, especially drought. We present a brief discussion of the precolonial land-and resource-use patterns, and then examine patterns, practices, and events of the past two hundred years. Though we briefly discuss contemporary conditions and trends, we focus on the period between roughly 1850 and 1950. This time period represents early settlement and resource exploitation practices, the more extensive Mormon colonization period of the late nineteenth century, and significant changes that occurred in the ensuing decades. Our analysis considers local experience as well as how local entities have negotiated water shortages in the evolving context of state and national laws and institutions.

Mormon settlement and agrarian practices are especially important to consider in examining conditions in the watershed today. These first Anglo-American settlements profoundly affected subsequent patterns of population distribution and resource use. The period of population expansion following initial Mormon settlement taxed the water resources available within the basin, and prompted development of local knowledge as well as adoption of successive technologies to address water scarcity problems. Cultural values that emphasized agrarian lifestyles' even while acknowledging that the region was far better suited for animal husbandry and grazing than for farming, strongly influenced the choice of strategies for coping with water resource challenges. Rather than adapting to grazing lifestyles typical of arid areas, residents built dams and introduced groundwater pumping as strategies for supporting crop cultivation. This approach still prevails today.

Post-World War II introduction of mining operations and power-generation facilities into the watershed, combined with rapid growth in tourism and recreation, compounded the stresses on water resources in the basin. By the late twentieth century, water demand throughout the basin combined with recurrent drought led to rising concerns about climate vulnerability. While recognition of local authority over water development and use persists, momentum has grown for introducing institutional mechanisms at the state and federal levels to increase adaptive capacity. Provision of scientific information about climate, hydrology, and resource management and promotion of structured planning processes are among the most prominent contemporary efforts to improve decision making at the watershed and local levels in the U.S. Southwest. In the process, the region's environmental history will be influenced as well.

## The Little Colorado River Basin: A Case Study

The evolution of economic, social, and cultural structures and dynamics in the Upper Little Colorado watershed reflects processes unfolding throughout the western United States. Occupied by indigenous peoples when the Spanish explorers first arrived in the early 1600s, southwestern landscapes and livelihoods were significantly transformed by Spanish introduction of livestock grazing and development of towns beginning in the mid-1600s, followed by expansion of farming activities with the arrival of Anglo-American settlers beginning in the mid-1800s and increasing after the end of the U.S. Civil War in 1865. Mormons wrought some of the most influential landscape transformations during this era through their extensive efforts to establish farms across the region, wherever growing seasons were long enough and sufficient water was available.

The arrival of railroad lines in the late 1800s produced another round of profound socioeconomic changes across the larger region, prompting rapid expansion of livestock herds, introduction of mining operations, and decimation of forests to provide energy for early mining and industrial needs. Cities such as Denver and Phoenix grew rapidly, especially at the end of World War II when families across the nation began moving westward in search of better lives. The growth of these urban areas, together with increased interest across a broad range of the U.S. population in natural-area tourism and recreation, introduced a new dynamic to rural areas: temporary—and often large—influxes of people during the prime months of summer (for hiking, fishing, etc.) and winter (for skiing and other winter sports), as well as growth in the number of cabins and homes owned by residents of the urban centres and, increasingly, by retirees.

Innovations in Internet technologies over the past decade and a half have allowed people to work in remote areas, further swelling the number of permanent residents in some towns. These changes, in turn, have fueled growth of rural towns and the number and variety of services offered in these towns. Meanwhile, across the entire time span from the arrival of the Spaniards to the present day, the native peoples of the western United States have sought to sustain their cultures and livelihoods, often under conditions of considerable adversity. They lost large portions of their previous domains and found themselves largely marginalized from the dynamics of the dominant society, yet their lands, practices, and experiences continue to stand as testaments to pre-European ways of knowing and being. These landscapes also testify to the profound disparities in social and economic well-being that exist among many tribal members. The Upper Little Colorado watershed reflects these conditions and dynamics.

### *Social Constructions and Biophysical Realities*

As historian Elliott West maintains, there is always a dissonance between how humans view the world and how they actually fit within it. Humans have the ability to imagine changes in the world even as they confront it: they visualize new connections and relationships that are not yet present (West 1998). The American West is one of the places where imagination has crucially influenced changes to the natural world, producing a series of imagined landscapes. Here, broad swathes of arid and semiarid landscapes have been transformed into irrigated agricultural landscapes and urban settlements. The Little

Colorado River Basin, whose modern settlement patterns grew out of Mormon visions of Zion and its related agrarian landscapes, is a microcosm of social constructions of the arid/semiarid West. The landscapes we see today, indeed, are the products of what historian Mark Fiege calls "vivid metaphor and compelling myths".

The Little Colorado River watershed provides a useful focal point for investigating the environmental history of relationships among climate, natural systems, and human systems: it is characterized by significant climatic and related environmental fluctuations, and it gathers into an identifiable geographical space a coherent narrative of human-environment relationships over time. The watershed, 26,964 square miles in size and ranging in elevation from some 4,200 feet to more than 7,700 feet above sea level, is located in the northeastern corner of Arizona. The area is characterized by topographic complexity, including elevational changes that produce spatial variability in temperature and precipitation and related ecological variability. While the area's climate is characterized by high levels of variability over decadal and longer time spans, climate-division data reveal long-term temperature patterns of cold winters and mild summers and a bimodal precipitation pattern of rather wet winters and summers, with dry spring and relatively dry fall seasons. These data show an average annual temperature for the area of 55[degrees]F, with average lows dropping to 7[degrees]F in winter (in the highest elevations) and ascending to highs of around 80[degrees]F in the summertime (Sheppard et al. 2002). It is notable, however, that over the past century overall temperatures have risen approximately 2.5 degrees. Though highly variable over time, precipitation averages approximately thirty-six inches a year at higher elevations, and about eight inches a year at the lowest elevations.

Scientific advances in the past several decades have improved understanding and predictive capability with regard to one of the major winter climatic influences on much of the Southwest: the El Nino-Southern Oscillation (ENSO). During El Nino years, the region tends to get wetter than average winters, while during La Nina years, winters are likely to be dry. Current forecasting capabilities allow skilful prediction of possible ENSO conditions as early as the fall preceding the winter rainy season; climatologists update these forecasts monthly based on measurements such as trends in sea surface temperatures in the tropical Pacific Ocean. The climate of the summer months, on the other hand, remains difficult to predict very far in advance. Unlike the broad-scale precipitation associated with winter storm fronts, summer climate conditions are characterized by highly localized, often intense thunderstorms associated with the North American Monsoon. As noted previously, the spring season tends to be very dry, but may include significant rain events triggered by synoptic-scale conditions such as tropical cyclones.

Paleoclimatic research highlights longer-term trends in the region's climate, including multi-decadal periods of drought alternating with periods of near-average or above-average precipitation. Such shifts have profoundly affected natural and human systems in the Little Colorado River Basin since the arrival of the first humans in the area.

Most of the land use in the basin is rural. Indian lands encompass almost half the territory, and only about 23 percent of the land is in private ownership. Forests of blue spruce, Douglas fir, and ponderosa pine dominate the higher elevations, interspersed with grasslands. Drier, lower-elevation areas of the watershed are characterized by sparsely vegetated desert scrub landscapes.

The Little Colorado River, with its tributaries, is the only significant source of surface water in the region. There are, however, many natural seeps and springs. The river, which is generally narrow and shallow and is fed by some thirty smaller tributaries, extends 250 miles between the White Mountains and the Black Mesa Highlands, eventually emptying into the Colorado River. Snowpack from the White Mountains is the primary source of water to the stream; in addition, water from the Coconino Aquifer, also flowing south from the White Mountains, surfaces as a series of springs along the main stem.

The water resources and vegetation coverage of the area are both influenced by climatic conditions, which are marked by high inter-annual to decadal and longer-term variability, as well as by geographical variability across the watershed. Winter precipitation and related snowpack conditions are crucial to sustaining both surface water and groundwater resources in the basin. Rain during the summer season also contributes to surface water availability although, due to the fast-flowing runoff produced by strong storms, summer precipitation is not a major contributor to groundwater aquifer recharge.

### *Paleo-environmental History of the Watershed*

Pre-Columbian Native American land-and water-use practices provide our earliest evidence of interactions between climate changes and corresponding cultural adaptations in the region. The Ancestral Pueblo, Hopi, and Navajo peoples all developed their own water management and shortage-mitigation strategies that, unlike drought adaptations in the nineteenth and twentieth centuries, hinged on mobility and community interconnectivity. Historians and archaeologists have pieced together a complex, evolving narrative of the earliest known habitations in northern Arizona, with most agreeing that humans entered the watershed between 30,000-11,000 BCE and 9500-9000 BCE. Archaeological evidence indicates that a mixed strategy of hunting, gathering, and horticulture prevailed by at least 600 BCE, and that in many areas, settlements were seasonal in nature. Responses to drought or flood events revolved around simply moving some distance away from the affected areas. With a shift toward greater agricultural complexity, populations grew, as did the use of available resources. The growth did not occur along a straight line, however.

An archaeological study of the Silver Creek watershed revealed significant population increases in the region around AD 900-950 and AD 1000-1050, but found a decline in population between AD 1050 and 1100. At least some of the fluctuation in demographic numbers seems to derive from a change in settlement patterns from small to large: the study found there was a population decline in small sites and an increase in large sites during AD 1100-1150 (Mills, Herr, and Van Keuren 1999). While scholarly disagreement persists regarding the temporal organization of increasing settlement complexity in the Southwest, most agree that by AD 1000 to 1150 human population growth and settlement densities had produced high levels of intercommunity interaction (Gumerman and Dean 1989). Relatedly, increased reliance on agriculture for subsistence increased local vulnerability to climatic variability, since the agrarian groups were less free to leave their crops and move to other areas than were hunter-gatherers. With reduced mobility, the Ancestral Pueblo peoples, for example, changed their practices to emphasize reliance on intercommunity interactions and mixed subsistence patterns (Gumerman and Dean 1989).

Communities along the Little Colorado River appear to have adopted a local system of shifting agriculture. For example, groups inhabiting the Homol'ovi area north and east of present-day Winslow, Arizona, shifted their agricultural activities in response to climate-influenced environmental conditions. As local population expanded and individuals were better able to support larger-scale irrigated agriculture, floodplain agriculture came to predominate along the Little Colorado River. During low river flows, however, these farmers were still able to grow crops using runoff and seepage fields, on stabilized sand dunes to the east of the floodplain (Lange 1998).

At the Ancestral Pueblo settlements in nearby Chaco Canyon (located in present-day northwestern New Mexico) agricultural and timber resource demands from a growing population severely taxed the ability of the environment to sustain a large population centre. A dry period between AD 1080 and 1100 probably caused many of the inhabitants to seriously consider the efficacy of unchecked demographic growth in the canyon (Lekson 2006). Agricultural choices would have been limited, for as population grew, ever more marginal land was brought under cultivation. With the advent of deep sustained drought in the twelfth and thirteenth centuries, the capacity of the land and water resources throughout the Colorado Plateau to support large, sedentary settlements diminished radically (Dean 1988).

Not much is known about activities in the region between the era of Ancestral Pueblo decline and the arrival of the Spaniards, though it is clear that the Spaniards changed the ecology of the region considerably through the introduction of livestock. Research suggests that the Navajo people entered the area in the fifteenth and sixteenth centuries, perhaps reflecting a strategy of migration from the Rocky Mountains of Colorado and New Mexico to escape the drought that occurred in that area during the late 1500s (Bailey 1980). Navajo subsistence practices revolved around hunting (by men) and agriculture (by women). Though some evidence of granaries suggests the existence of surpluses, trading between settlements was likely to occur only when climatic conditions favoured sufficient surplus production (Brooks 2002). The introduction of Spanish livestock from Spanish settlements on the upper Rio Grande in the seventeenth century resulted in adoption by the Navajo of sheepherding; the demands of pastoralism, in turn, led to a dramatic expansion in the people's use of the land and its resources, as well as to social stratification and military aggressiveness (Bonte and Galaty 1991). At the same time, pastoralism also concentrated people and animals into the areas where water and forage were in greatest abundance (West 1998). As resources were depleted, Navajo residents migrated west and south to new lands, leading to pressures on the neighbouring Hopi and Apache populations (Mauss 2003). The drought of 1736-48 prompted yet another southwesterly migration (Brooks 2002).

## Euro-American Arrival, Mormon Settlement, and New Environmental Stresses

Although Spanish expeditions entered the area in the late 1500s, the Upper Little Colorado watershed saw little non-native settlement until the arrival of the first Mormon settlers in 1876. These individuals brought with them an ethos of sedentary agriculture based on irrigation technologies and corresponding religious institutions that, among other things, determined settlement distribution, influenced land-and water-use practices, and provided a framework for dealing with drought.

The arrival of Mormon settlers was preceded by exploratory trips undertaken by Mormon Church members between the 1850s and the early 1870s. Church leaders, including Brigham Young himself, found the area to be acceptable for Mormon expansion and redemption of desert lands and, equally importantly, believed that the move would facilitate religious conversion of the area's Hopi and Navajo inhabitants. Young commanded specific groups, chosen for their devotion to the faith and their loyalty, to move from Utah to the Little Colorado watershed and establish self-sufficient communal orders (including communal ownership of water) that would be based on irrigated agriculture.

The settlers arrived at an auspicious time, preceded as it was by two five-year periods of wetter than average conditions: the period from 1833 to 1842, with a total of eighty-four inches of precipitation, was the second wettest decade on record in the region; the years from 1865 to 1869, with almost forty-four inches of total precipitation, constituted the third wettest five-year period. It was during the latter period that the Mormon Church sent the series of exploration parties to the watershed. Given precipitation conditions in the years before their arrival, it is perhaps not surprising that their reports depicted the area in favourable terms. It is interesting, however, that the settlers viewed the many Indian ruins in the area not as a warning about the dangers of overtaxing the environment with excessive human occupation, but as a sign that through their religious work the land could be resettled: "to our settling this new land, we had evidence that this land has been densly [sic] populated, this choice land will be blessed according to our faithfulness" (J. Smith 1881).

The reports encouraged Mormons to migrate to the area but also generated unrealistic expectations with regard to the region's agricultural capacity. Ironically, in their efforts to protect their religious beliefs and practices and avert competition over land and ideology (Journal of Discourses 1862), Mormons often sought out lands deemed undesirable by other people. In the Little Colorado watershed, as discussed in more detail later, realization of the full extent of the limits of the land and its resources emerged all too soon with the return of deep and long-lasting drought conditions in the 1890s.

Throughout the colony, in local church and community meetings, the settlers talked of trials from God and the need to improve the land through communal endeavor and thus spread Mormonism to as many lands as possible (Leone 1974, 1979). Limiting the colony's success, however, was their philosophy, which focused not on how much rain fell or how much water was in the river, but on how hard they worked at improving the land.

The Mormons in this way were like many other western settlers who believed that a religious work ethic would transform arid land into agricultural paradise (H. Smith 1950). This logic dictated that through producing and sustaining a thriving agricultural community, they would achieve victory over natural forces. The settlements in Utah had proven that victory was possible: in 1879 Wilford Woodruff, future president of the Church, "spoke of the first settlement of Utah, that it was a desert, the country is now full, and we have not got room. Zion has to lengthen its cords, it matters not where we are.... It is our calling to build up the Kingdom of God" (Eastern Arizona Stake 1979-1886).

The topography of the region, which includes steep slopes and elevational changes that influence local ecological forms and processes, makes agriculture challenging at the best of times and under the best of circumstances. At the time of Mormon settlement, the best agricultural areas lay in the southern portion of the watershed; here water was most abundant and flooding was less intense. However, although many joined Jacob Hamblin in recognizing that the ecology of the watershed was, overall, best suited for grazing—sheep grazing was the primary land use in the area when the Mormons arrived—most Mormons acted on their desire to plant grains and thus improve the land to a "higher" form of land use. As one settler put it, "our people are an agricultural people and wish to beautify and adorn their homes and to improve and cultivate the land. I believe that the Lord has selected this new land for his Saints to gather to, as in thickly populated places wickedness is more prevalent.... In these new places there is a feeling of grandeur, a feeling that brings God nearer to us, and brings us nearer to Him" (Snow, quoted in J. Smith 1881). Creating this Kingdom of God required building dams and water-diversion projects, which would eventually render much of the river channel near the settlements devoid of riparian vegetation.

Upon their arrival in the basin, Mormon settlers identified two main areas for community development. One was in the middle of the river's length, in an area stretching from modern-day Winslow to the town of St. Johns, where a few Mexican settlers had already established homes near a post built by Solomon Barth to supply soldiers stationed at Fort Apache. These early settlers, who herded sheep and grew a little grain (Fish 1895[?]), were most likely immigrants from New Mexico who were attracted to the supply post. A small settlement also existed on Silver Creek, near present-day Snowflake, when Henry Stinson arrived and began building irrigation ditches there in the early 1870s (Fish 1895[?]).

The first Mormon settlers had little knowledge about the Little Colorado River or its flow regime, or about the ecological and climatic processes of the region. However, over the last quarter of the nineteenth century and first decades of the twentieth century, they built, in their church meetings, a body of local knowledge and series of strategies for responding to climatic stress. Their approach, which blended observations of local conditions with religious doctrine, fostered community cohesiveness at the same time that it sought to reduce vulnerability to environmental stresses. As settler Fredrick Christensen observed, "It [is] nescessary [sic] to learn by experience hence we should school ourselves and be prepared to return back to the presence of our Father. Let us stick to this work, for this is a preparation to go back and redeem the Centre Stake of Zion" (St. Joseph Stake Minutes, 1878).

The irrigation-based agricultural livelihood practices of the Mormons, especially during the early years before the introduction of water pumps and sources of energy to run them, dictated a settlement distribution pattern that was largely based on proximity to the river and the ability to take advantage of gravity to direct flows into the fields. Settlers' decisions about where and how to plant were also influenced by ecological conditions that had been produced by the previous series of wet years. Yet, with the onset of deep drought in the 1890s, much of the knowledge that had informed planting decisions had to be rethought. Lack of sufficient reservoir storage limited the range of options available to some farmers, while sporadic flood conditions forced others to deal with damage to farmlands and structures created by dam failures.

The spatial and temporal variability of stresses on the communities in the watershed reflected to a large extent, then as today, spatial and temporal variability in climatic and hydrologic conditions. This variability helps account for the fact that some of the Mormon settlements established with high expectations did not survive past the nineteenth century. The settlements of old Taylor, Obed City, and Brigham City were among the earliest communities established, having been built by the first group of migrants arriving in 1876, yet all were virtually abandoned before the beginning of the twentieth century. Obed, located amid wetlands, dissolved in the wake of a malaria outbreak in 1877 (Jensen n.d.). At old Taylor and Brigham City, repeated dam failures and resulting crop failures led to the communities' demise. Woodruff often borrowed food supplies and manpower from other settlements, due to its exposure to spring and late-summer floods: the settlement was located below the confluence of a seasonal tributary with the Little Colorado River. Between 1876 and 1890, the settlement experienced ten dam failures. Thirteen dam failures occurred near the settlement of St. Joseph over the same period.

Archival information indicates that these downstream communities have always been more susceptible to flooding and water shortages than their upstream neighbours. People living in the downstream stretches of the river had to depend on surface water that was infamous for its unpredictability and its heavy silt load. These communities were eventually forced to resort to groundwater to overcome problems caused by irrigation activities on the main channel of the Little Colorado River. Dissatisfaction with communal institutions arose during the later decades of the nineteenth century, eventually resulting in their abandonment. The demise of the communal culture occurred in no small part because of the difficulties in sustaining lives and livelihoods in the region (Peterson 1986).

More broadly, though some settlements were successful and remain thriving communities today, the early Mormon experiences reveal the types of problems that arise when significant disparities exist between values and livelihood practices on one hand and inescapable environmental limits on the other. Many settlers, upon abandoning communities in the lower basin, moved upstream to establish new towns such as Snowflake and St. Johns. Here they found greater success, for they faced much less difficulty in developing surface water and groundwater resources. The settlement of Snowflake, for example, purchased from Henry Stinson in 1878, quickly became the most productive agricultural settlement in the region because of its relatively rich water resources. Silver Spring and Silver Creek were—and still are—the most consistent water sources in the basin. Rather than fluctuations in water availability, the primary challenge for the upper-basin settlements such as Springerville, Eager, Alpine, and Pleasant Valley was the short growing season due to higher elevations.

Yet even in these communities, concerns about water shortage persisted. In 1886, for example, the Joseph City Irrigation Company mandated a fee of $7 per acre for water used by its members. The revenues were to be used for building and maintaining water infrastructure (Mien City Minutes 1876-1886). These fees continued to rise as St. Joseph's water problems persisted: there was insufficient water when they needed it in the spring, and too much water when they did not need it in the late summer. By 1894, water fees had risen to $20 per acre, forcing individuals to plant more acreage.

With the arrival of the railroad in 1881, communities had access to more sophisticated irrigation and water-diversion technologies. At the same time, however, the railroad brought large-scale cattle grazing into the watershed (Arizona Gazette, Feb. 26, 1915). The relocation from Texas of the Aztec Land and Cattle Company (also known as the Hashknife Outfit), with herds of as many as 40,000 cattle, was particularly instrumental in driving environmental change. Mormon settlers strongly objected to the cattle operators' encroachment on and claiming of vast tracts of land, noting a lack of respect for the rights of others (LeRoy and Wilhelm 1982). The issue of encroachment was especially critical to the Mormons, for legal practices at that time persistently failed to officially recognize Mormon settlement claims.

A series of wet years in the late 1880s, coinciding with the arrival of the Hashknife outfit, prompted the group to ship in even more livestock and eventually to run 150,000 head of cattle and 120,000 head of sheep in the area. These animals shared the range with Mormon herds and with greatly increasing numbers of Navajo sheep, thus placing enormous stress on the area's rangelands. Conditions changed, however, due to deep drought beginning in the mid-1890s and extending to 1904. During this time period, the area received less than half the precipitation that had fallen during the previous wettest five-year periods. During the five-year period in the depth of the drought, 1900-1904, total precipitation amounted to a little less than twenty inches (Ni et al. 2002). The environmental and social impacts were huge. Mormon residents commented that this drought had "no parallel in recorded history" (LeRoy and Wilhelm 1982); water was scarce, and the grass disappeared. One settler observed, "It is stated that three fourths of all the stock along the river will die. The stock men feel about ruined and the outlook looks dark for all classes of business" (Fish 1895[?]). Widespread erosion and siltation of waterways were among the environmental effects of the interactions between drought and overgrazing. The comparison with conditions when the Mormons first arrived was stark:

When we came to Arizona in 1876, the hills and plains were covered with high grasses and the country was not cut up with ravines and gullies as it is now. This has been brought about by overstocking the ranges. On the Little Colorado we could cut hay for miles and miles in every direction.... Water followed the cattle trails and cut the country up. Later tens of thousands of cattle died because of drouth and lack of feed and disease." (quoted in Tellman, Yarde, and Wallace 1997).

A study of arroyo-cutting in the Southwest in the late nineteenth and early twentieth centuries (correlating roughly with the years of Mormon settlement) found widespread erosion throughout the region. During this time, the Southwest was characterized by rapid and pronounced erosion as well as by the desiccation of river valleys. Social factors limited locally to the watershed, such as Mormon irrigation practices and cattle grazing by the Hashknife, constituted one but not the only driver of erosion (Cooke and Reeves, 1976). Indeed, erosion and arroyo cutting, especially on the Little Colorado, is best explained as the outcome of human activity combined with climatic variability and changes in vegetation cover. Maitland Bradfield (1971), in his study of Hopi agriculture, found that climatic variability and loss of vegetation contributed heavily to erosion in the Little Colorado River watershed between 1880 and 1900, 1971). He notes, however, that late-nineteenth-century arroyo cutting along the Little Colorado River and elsewhere can only be explained by combining the impacts of social factors with the impacts of climatic variability. Clearly, though, the 1890s drought contributed substantially to erosion processes as well as to corresponding water-resource problems throughout the Mormon settlements.

## Twentieth-century Institutional and Technological Responses to Climate Variability

The impacts of the turn-of-the century drought on water resources prompted significant institutional change, including the first systematic effort within the basin to record the water rights of each holder. This effort foreshadowed modern water-rights institutions that organize rights to water chronologically, with the claimant having the longest history being accorded first priority, and so on. The information produced by the recording process prompted a change in water allocation practices in times of shortage: rather than relying on community decision making, water was henceforth allocated based on individual recorded rights. For instance, in Joseph City, on April 15, 1907, a committee of three was appointed to "record the dams and water according to their judgement" (Porter n.d.) The Joseph City Irrigation Company, previously responsible only for construction and maintenance of dams and irrigation ditches, was now charged with managing water allocations in response to drought-induced shortages.

These sorts of institutional changes also afforded opportunities for coping with variability within long-term droughts. For example, while 1904 stands as one of the driest years on record (with a mere 1.6 inches of precipitation), 1905 (at 13.7 inches) was the third wettest year since the late 1890s. Such wide swings in inter-annual variability made it virtually impossible to sustain consistent agricultural output by relying solely on surface flows. The first strategy adopted by the settlers for coping with variability in water availability was impoundment of water in reservoirs. The St. Johns Stake, for example, built seventeen dams between 1883 and 1921 on the main stem of the Little Colorado River and its tributaries (LeRoy and Wilhelm 1982). Dams and reservoirs posed their own challenges, however: violent summer storms could produce large volumes of water that led to breaches in the clams and consequent loss of life and property, while extended drought conditions diminished the amount of water entering the reservoirs.

Around the same period, the Joseph City Irrigation Company made the first serious attempt to systematically explore for and exploit groundwater resources (Porter n.d.). This effort notwithstanding, groundwater development in much of the basin remained minimal until the mid-1900s. As late as 1946, only 600 acre-feet of groundwater were being pumped in the Joseph City-Holbrook area (Babcock and Snyder 1947). By contrast, groundwater pumping in Navajo County was much more intensive, producing an estimated 11,000 acre-feet around 1950 and more than 38,000 acre-feet by 1972. The growth in groundwater development coincided with three of the driest ten-year periods in the instrumental record for the area: 1942-1951, 1955-1964, and 1968-1977. During these decadal periods, precipitation amounted to only 73, 82, and 83 percent of average.

Groundwater provided distinct advantages over surface water: it was available year-round and was less sensitive to climate variability (Abruzzi 1985). It also facilitated economic growth and development in the basin, including construction of three coal-powered generating stations (near St Johns, Springerville, and Joseph City), a paper mill, and a mining operation. These industrial activities increased demand pressures on available water resources and generated tensions—which still persist—among local residents. Increased appropriation of water combined with drought conditions prompted the state of Arizona to declare a "critical groundwater area" near Joseph City in 1948. The designation, formalized as the Joseph City Irrigation Non-Expansion Area in the state's 1980 Groundwater Management Act, was introduced as an effort to curtail declines in groundwater supplies through prohibiting expansion of irrigated acreage. The groundwater basin underlying the Joseph City portion of the watershed came under particularly intense pressure beginning in 1960 due to substantial growth in water demand emanating from construction and expansion of the Cholla power generating plant. The Arizona Public Service (APS) utility, which serves the greater Phoenix area, had bought land or water rights within a 16.5-mile radius of the plant; its purchases included all of the wells except two that were owned by the Joseph City Irrigation Company. These purchases plus designation of the groundwater basin as the Joseph City INA served to preclude further competition between irrigation and power plant water uses. Today, use of water for irrigation continues to be restricted to acreage that had been irrigated during the years from 1975 through 1979 (ADWR 2006). The amount of water pumped by industry is not insignificant: the Cholla generating station used around 7,425 acre-feet of water in 1980; the paper and pulp mill constructed near Snowflake was using 16,469 acre-feet in the same year. At the same time, however, the amount of water extracted for agricultural uses declined in tandem with reduction in the amount of farmland being actively irrigated. Today, although irrigated agriculture continues to use the greatest amount of water in the region, irrigated acreage has decreased. For example, in Navajo County, irrigated farmland decreased from 11,390 acres in 1978 to 6,351 acres in 1992. Nevertheless, agriculture today accounts for less than 2 percent of the region's total employment, a sharp contrast to the 40 percent thus employed in 1940. The fastest growing water use, by contrast, is in industrial and export activities (U.S. Census Bureau 2000).

Whereas the drought of 1955-1964 and dry conditions between 1968 and 1977 (Ni et al. 2002) placed pressures on the growing and sometimes conflicting water demands within the Little Colorado River watershed, the wettest conditions in the instrumental record occurred on the heels of the latter drought period. It was during this period, beginning in 1978, that adjudication of water rights began within the watershed (ADWR 1989). It is notable that the 1980 Groundwater Management Act was enacted during this period. The act emerged in response to federal demands that the state get its groundwater overdraft problems under control before approval would be granted for construction of the Central Arizona Project (CAP), which would deliver Colorado River water to the state's main cities and the heavily agricultural area in Pinal County (located between Phoenix and Tucson). While strong regulation of groundwater applied only to the central corridor of the state, the act did address rural issues to a minor degree with establishment of state-level administration, through the Arizona Department of Water Resources (ADWR), of surface-water rights as well as responsibility for collecting and archiving data about water resources in the state and regulating the spacing of groundwater wells.

A proposed interbasin transfer to support mining-related operations outside the Little Colorado watershed illustrates one hotly contested aspect of contemporary water politics in Arizona. In a 1992 decision, the special master of the Arizona General Stream Adjudication addressed a dispute over a proposal by Phelps Dodge to divert water to which it had rights in Show Low Lake to its operations in the Salt River and Gila River watersheds. The special master mandated that "when water is diverted into another watershed or river system as part of an exchange agreement, DWR shall report information on all aspects of the exchange sufficient to allow potential objectors in the watershed of origin to make an informed decision about whether or not to object to the claimed water rights" (Office of the Special Master 2003). The decision illustrates the limitations of state authority over local water resource management: DWR is required to supply information, but the actual decision on whether a water transfer can occur is left to the parties involved. In this case, the dispute became moot when, in 2005, Phelps Dodge Corporation relinquished its rights to Show Low Lake.

## A Contemporary Environmental History of the Watershed

Access to—and wise use of—water resources is key to sustainable habitation in the Upper Little Colorado watershed. Climatic variability and change, combined with significant changes in population dynamics and related resource use over the past several decades, underlie water resource vulnerability in the area. Significant growth in tourism and recreation activities, for example, draws large numbers of people to the area. Though the area is experiencing greater winter tourism and recreation such as siding, the area continues to host considerable numbers of visitors during spring, summer, and fall, when water resources may be most stressed. Also significant have been negotiations involving settlement of Indian water rights claims. Climate factors, including a marked warming trend over the past several decades, wet conditions between the 1980s and mid-1990s, and reemergence of drought in the late 1990s and early 2000s, continue to pose challenges to efficient and equitable management of the region's limited water resources. Population growth remained high in the most recent census, showing increases of 4 percent in Winslow, 12 percent in Pinetop-Lakeside, and 22 percent in Show Low between 2000 and 2004. During roughly the same period, between 1999 and 2003, only twenty-three inches of precipitation fell in northeastern Arizona (Morrill 2005). A comparison of this drought with the extended drought period from 1947 to 1956 indicates that, while both periods exhibited significantly below-average precipitation, the latest drought resulted in a consistently lower snowpack. In addition, warmer temperatures combined with the timing of snowfall affected runoff and recharge patterns (Morrill 2005). Research into these conditions suggests that global warming might be implicated in producing more intense drought conditions, in which case greater problems are likely in the future.

Deep, sustained drought exacerbated by climate change may well constitute the gravest threat yet to the sustainability of the human and natural systems in the watershed and beyond. This threat raises serious questions about the future of the area's water resources and demands ongoing adaptation, including identification and enactment of measures to avert or mitigate the impacts of climate variability and change. Native peoples within the basin have the longest history of interaction with the biophysical environment; however, it is only in the past few decades that they have begun to recapture control over certain important portions of their resource base. Native peoples, in contrast to the other cultural groups in the watershed, place a high spiritual value on water bodies and water resources, a value that embodies the unique environmental histories of the different groups. In the case of the Zuni people, recent developments have led to successful reappropriation of an important water source. The Zuni people had for centuries inhabited land along the Little Colorado and Zuni Rivers, but by 1877, they had lost all but 3 percent of their traditional lands. Most serious was a federal executive order that forced tribal members onto a 408,000-acre reservation in western New Mexico. This forced migration deprived the Zunis of the right to live near the Colorado River, thus disrupting their spiritual practices. Since at least the sixteenth century, the Zuni people have traveled to a sacred location known as Zuni Heaven, a lake located near the confluence of the Little Colorado and Zuni Rivers that has now dried up due to upstream water diversions and groundwater extractions (Committee on Indian Affairs 2003).

Years of effort to reacquire the lake finally culminated, in 1984, in establishment of the Zuni Heaven Reservation for the primary purpose of furthering "the religious and cultural needs of the Tribe" (Committee on Indian Affairs 2002). Once again in possession of their spiritual lands, the tribe began restoring the riparian areas around the lake; in 1990, Congress contributed to their efforts through establishment of the Zuni Heaven Resource Development Trust Fund, which provided $25 million to help restore the area's wetlands. Further assistance materialized in the Water Rights Settlement Agreement of 2003, which empowered the tribe, for purposes of restoring the wetlands, to purchase up to 3,600 acre-feet of surface water rights from upstream sellers and to pump up to 1,500 acre-feet per year of groundwater (Committee on Indian Affairs 2003). Mormon settlers relied on a different set of spiritual values to link their communities with the land and its resources. For example, frequent dam failures and fluctuations in the length of growing seasons, along with the resulting variability in crop output, prompted reliance on religious tenets that viewed natural events such as flooding as trials or tests sent by God and adherence to notions of communal work as religious ritual to improve upon nature. The Mormons' religion-based decision making in the face of environmental stress often involved negotiations between communities that were competing for the same resources. For example, residents of Joseph City (then called St. Joseph) invoked notions of religious community to compel upstream residents of Woodruff to breach their dam. Even though this act could have compromised their ability to grow their crops, Woodruff complied, though they observed that "after working so hard and under such adverse circumstances this was pretty severe medicine" (Dean n.d.). The early reliance on religious rhetoric has since faded, and Mormons no longer make up the dominant Euro-American presence in the watershed. They do, however, continue to represent a locally strong presence in communities such as Snowflake and Joseph City; in locations such as these, religion continues to influence resource management decisions.

The history of the Little Colorado River Basin is one of oscillating periods of plenty and scarcity, with climate constituting a significant but far from determinant influence over the distribution of vulnerabilities to floods and drought. Climate-society interactions in the watershed have strongly influenced the area's biophysical and human environments and have prompted successive innovation and revision of institutional mechanisms aimed at sustaining lives and livelihoods. For the early Mormon settlers, decision making resided at the community level where the main concerns were sustaining the settlements through agricultural production.

Organization of the settlements into stakes, while primarily serving religious and community-building functions, also served as venues for addressing problems generated by environmental variability and change, including variability in hydrologic processes and water availability. Successive droughts, however, eventually resulted in a transfer of decision making from the community at large to irrigation companies, which focused more on rational water management practices than on religious ideals. Throughout the first half of the twentieth century, these companies increasingly turned to groundwater to buffer drought-induced variability in surface-water availability and to meet the considerable growth in water demand. Increases in groundwater use, however, produced overdraft conditions in the area's aquifers, most notably those in the Joseph City area. Here, water table decline induced conflict between power-generation and agricultural uses, eventually leading—for the first time—to substantive state-level intervention and regulation through creation of the Joseph

City Irrigation Non-Expansion Area. The Little Colorado watershed faces an uncertain future characterized by growth in the number of groups wing for access to water and the possibility of greater climatic variability, long-term warming trends, and deeper and longer-lasting droughts. Under such conditions, state-level regulation is likely to increase. A report summarizing an evaluation of the ADWR by the Arizona auditor general has already called for state-level regulation of water throughout the state (Norton 1999).

Knowledge of the environmental history of the watershed provides valuable insight into the experiences, values, knowledge, and practices that have produced the conditions existing today. These "thick" narratives (Geertz 1973), if integrated into development of plans, decision processes, and decision tools, hold promise for improving citizens' acceptance of and support for policies aimed at ensuring the sustainability of the watershed's closely coupled human and natural systems. Scientific research aimed at understanding the nature, intensity, and potential impacts of climate variability and change, and at developing products designed to support policy and decision processes, can also benefit from understanding the contexts in which the products and knowledge will be used.

# Chapter 7

# Water Diversion Management

Water is increasingly a major political issue as scarcity of the resource grips several societies, especially in developing countries where agriculture is responsible for about 80 percent of water consumptive use. However, western nations are not immune to water tensions, for instance in Greece, Spain, Italy, or the western United States, where the available water is being exploited to the limit: the Colorado no longer reaches the sea, and growing debates are emerging as to whether water should be allocated to thirsty cities or to agriculture; whether public funds should be invested again to increase the resource or demand management implemented; and whether water could be imported from far away. These questions are increasingly relevant: consumption patterns of water in the western US are clearly not sustainable. Given the technology available today, massive water transfers could only come from Canada. There have been several mooted projects in this vein, mostly in the 1960s and 1970s. Should Canadians worry about water exports to the United Sates, especially in the frame of NAFTA? Or is climate change going to be the decisive factor in the debate?

## The Advent of Water Addiction

Water is a key ingredient in the fabric of the western American society, as has been well studied by Donald Worster. The west is not completely waterscarce, for several mighty rivers flow in the region, mainly fed by snow and glaciers from the Rockies; but it definitely is a semi-arid region, compelling all societies living there before the industrial revolution to adapt to water scarcity. Early 20th century American society, empowered by the industrial age, decided to harness rivers and aquifers. Technology enabled American society to eliminate the water scarcity burden and developed the illusion that technology would always bring about a solution to growing water needs. "What nature does not yield freely, humanity should refashion to better suit human needs.... Nature has no greater purpose than to serve Man, and Man has no greater purpose than to work the land and take his place in the productive cycle."

Wendy Nelson Espeland also clearly depicted the representation that developed at the time that all water flowing unused to the ocean was a wasted resource, and that rivers needed to be "tamed" and "harnessed" to be put to use. President Franklin Roosevelt declared in 1935, when inaugurating Boulder Dam, that "the mighty waters of the Colorado were running unused to the sea. Today we translate them into a great national possession." This idea is still very much alive today. "Much of the contemporary culture of the American Southwest is based on denying its desertness.... In a subdivision being built to the south [of Las Vegas], Paseo Verde Parkway and Val Verde Road intersect in Green Valley Ranch. The concept of green, like sod lawns, was an imported fantasy."

Technology can make up for water scarcity and water must be used so as to develop western resources: "The rural Western ethic is that all wealth comes out of the ground, either as grass growing or as minerals being mined.... So the fact that today's reclamation projects-such as Garrison, CUP, Animas-La Plata-cost a few million dollars for each farmer they put on the land, doesn't cause their proponents to blink. That, they say, is the price society pays for creating the stuff of wealth. Without it and the other industries based on earth, there is nothing."

## Early Massive Water Transfer Projects

Massive water transfers were first built in the eastern United States in 1847 with the Croton aqueduct for New York City and the 1900 Chicago diversion. As early as 1906, water transfers began being built in the west, with a 46 km-long canal from the St. Mary River in Montana to the Milk River, triggering a severe dispute with Canada that forced the United States to negotiate the 1909 boundary waters treaty. Other projects soon followed: the Los Angeles aqueduct in 1913, the Hetch Hetchy water supply (1934), the Colorado River aqueduct (1941), the Ail-American Canal (1942), and the California central valley project (1951). Diversions were initially designed for municipal water supply, but irrigation soon became their main rationale, as the sector benefited from generous subsidies from the federal government through the Bureau of Reclamation (created in

1902), its agency dedicated to the development of agriculture in the west. When exploited resources began to show signs of exhaustion, engineering firms started to design huge transfer projects from Canada. But the main trigger of the idea that water resources were exhaustible and not adequate to sustain limitless development in the western United States struck in 1963 when the supreme court compelled California to stop excessive Colorado water pumping. Colorado River water had been apportioned between upper-basin and lower-basin states with the Colorado compact of 1922, but Arizona felt California tried to prevent its own use of the water. With the 1963 ruling, California's share was established at 4,4 million acrefeet (af), by far the largest share in the 15 million af available, but a figure that set a limit to what California could bank on. The very idea that the water resource was all of a sudden set at a strict limit sent shockwaves throughout California that triggered a frenzy of plans designed to increase supply through continental water diversion projects from Canada.

These projects, like the earlier water diversion schemes, involved diverting large volumes of water over long distances, but their sheer size placed them on a very different scale, both regarding their length (often greater than several thousand km) and their cost (several dozens of billions of dollars). They were designed, from 1959 to 1985, in the United States and sometimes in Canada, on a magnitude no project had ever reached before. These projects involved the transfers of the Mackenzie River, for instance, all the way through the flooding of the Rockies Valley to the Mississippi and Colorado Rivers, or the diversion of the Great Lakes, or even the damming of James Bay so as to convert it into a freshwater lake and its subsequent transfer to the American west. Costs involved reached hundreds of billion dollars and environmental impacts were not studied at the time.

## Massive Continental Water Diversion Projects Never Implemented

Although these continental water diversion projects received considerable media and state government attention in the west in the 19605, none went beyond the feasibility study level and none managed to get an official approval.

### *Cost*

Water is heavy. When transported by aqueducts over long distances, it requires a lot of energy merely to be moved and pumped over obstacles. The Grand Canal, designed in 1959 by Canadian engineer Tom Kieras would have required at least 6 nuclear plants merely to pump the water from James Bay to Georgian Bay in Lake Huron. It is therefore expensive to operate these infrastructures. Moreover, the capital requirement is huge, as attested by their estimated construction costs at the time of the project design. Governments were more than reluctant to invest such large amounts of money for water exports, especially when the profitability of such projects was far from certain. Estimations of the cost of water transported or produced by different means show that desalination is becoming much cheaper than it used to be: about \$0.6 to \$1.2 per cubic meter, as compared to \$0.55 to \$1.35 for plastic container transportation; \$1.25 to \$1.5 with water-carrying ships; \$0.6 to \$0.85 with iceberg transportation; \$0.07 to \$1.8 for water recycling; and \$0.8 to \$3 for transfer canals of about 500 km. It is true these means do not all serve the same purposes, as desalination plants are usually located by the sea and serve urban demand. Sea water desalting technology, in particular, improved so fast between 1985 and the present that operation costs have been reduced dramatically and divided by five, from \$2.5 to \$0.55 per cubic meter.

Desalting is now a very affordable water-producing technology for urban and industrial consumers, enabling them to tap into an inexhaustible source. However, the water it produces cannot be considered for irrigation purposes, given its cost, its distance from the interior, and the large volumes irrigation demands. But these figures are useful inasmuch as they show that water transfer by canals is not as inexpensive as many advocates insist it is: it is not certain that farmers could afford their expensive water volumes without massive subsidies.

### *Stabilizing Demand in the United States*

The main reason that the giant diversion schemes never came to fruition is that demand was not really present in the United States for Canadian water. Several factors explain why these projects lost their allure for both governments and society. The budget crisis that began in the early 19705 in the United States precluded public money for such large endeavours, at a time when public opinion gradually turned against large scale engineering projects that tampered with the environment: 1977 was the "hit list" year when President Jimmy Carter discarded several water development projects in the west because of their extremely low benefit/cost ratio. The economics of continental diversion projects have so far worked against them and will do so for several more years.

It was all the more difficult to argue for these continental diversion schemes as water use, contrary to what had been expected in the 1960s, showed a definite trend toward stabilization, especially in agriculture. More cost-recovery-oriented water pricing; competition from other regions, mainly Asia; and cost incentives that lure American producers to Mexico are among the factors that explain why water use in agriculture remained roughly stable between 1975 and 2000 throughout the country.

If the federal government does agree, during the Doha round of trade negotiations, to reduce agricultural subsidies, water prices for farmers could increase markedly, resulting in financial incentives to consume less. Alternatively, water demand for irrigation could decrease as the result of farmers going out of business. As a whole, water withdrawals in the US have increased slowly, much more slowly now than its population, and could even begin a downward trend should competition from foreign fruit and vegetable producers work against local farmers.

In addition, although there still is room for improvement, water use per person is showing signs of stabilization, probably thanks to both tariff and education policies. The evolution of water use in the United States, from 1970 to 2000, shows that although the population grew by 38.6 percent, water use increased by only 10.7 percent. Urban consumers began to value water conservation with the development of urban water use reduction programs in most large cities. Industrial use actually decreased by 55 percent, whereas irrigation grew by only 6.1 percent.

Therefore, although water is still used at an unsustainable rate in the western United States, importing water from Canada is not as urgent as appeared to be to some politicians a few years ago, such as former senator Paul Simon. A stabilizing trend in water withdrawals and the availability of cheaper sources with desalting plants led public planners to forget about massive water transfers from Canada. When asked, even the Western States Governors Association estimated continental diversion projects were unlikely to ever be built."

### Opposition from Within the United States

Moreover, there also was opposition from within. Governmental archives from the early 1980s attest to the western United States lobbying for the diversion of Great Lakes water to quench its thirst. The International Joint Commission, created to prevent and resolve disputes between the United States and Canada under the 1909 boundary waters treaty, explicitly warned against water diversions from the Great Lakes basin in its 2000 "Final report on protection of the waters of the Great Lakes."

Great Lakes states wanted to resist these projects, both for environmental and political reasons: why would the Great Lakes states give California added value at a time when so many firms were leaving the area and moving to the West Coast? The Council of Great Lakes Governors (CGLG), created in 1983, is a partnership of the governors of the eight Great Lakes states and the two Canadian provinces of Ontario and Québec. The Great Lakes charter stemmed from their growing concern that Great Lakes water could be diverted to water-scarce regions of the United States.

The charter, signed in 1985 by the CGLG members, created a notice and consultation process for Great Lakes diversions. The signatories agreed that no Great Lakes state or province would proceed with any new or increased diversion or consumptive use of Great Lakes water over five million gallons per day without notifying, consulting, and seeking the consent of all affected Great Lakes states and provinces.

After an Ontario firm, Nova Group, tried to tap water in 1998 from the Great Lakes for export purposes, triggering an uproar of protests from both the Canadian and the American sides of the lakes, the CGLG decided to further strengthen the water export ban. This led to the Great Lakes charter annex, signed in June 2001.

The annex outlines a series of principles for reviewing water withdrawals from the Great Lakes basin that is grounded in protecting, conserving, restoring, and improving the Great Lakes ecosystem. In December 2005, the annex 2001 implementing agreement was signed, with specific guidelines for the enforcement of the principles.

It will also form a legal basis for the ban when the Great Lakes-St. Lawrence River basin water resources compact is ratified by the US congress, making it legally binding for American Great Lakes states. With this agreement, water diversions are only permitted over short distances from the Great Lakes, must ensure that water is returned to the Great Lakes basin, and are only permitted if there is no other solution to satisfy the water demand.

## Massive Transfers

Canadians became very fearful in the late 19905 about potential American water transfer schemes, leading to calls for the federal government to enact a ban on water exports. Wisely, Ottawa refused to enact any law that would implicitly recognize water as a good, but rather put forward bill C-6, voted in December 2002, which forbids interbasin transfers. Under NAFTA, this was a clever move inasmuch as it leaves water outside the realm of commerce and applies to both Canada and the United States. But Canadians were ignoring the fact that large-scale water transfers already exist throughout the world, and especially in Canada. Several massive transfer schemes have already taken place: in British Columbia (the Kemano transfer, 1954); in Manitoba (the Churchill derivation, 1976); in Ontario (the Long Lake and Okogi River transfers, 1939 and 1943); in Labrador (the Churchill Falls transfers, 1971); and in Québec (the Caniapiscau and Eastmain transfers, 1985).

There are major differences from American water diversion projects: Canadian transfer schemes were for hydroelectric development purposes, and not for urban consumption or irrigation. They are usually massive and over short distances, less than 200 km, whereas American water diversions often run for more than 500 km. There are currently plans to divert the Peace River in northern Alberta to meet growing agricultural and urban water needs, but that project is far from accepted. In addition, HydroQuebec is currently developing other water diversion projects for the Betsiamites, Eastmain, and Manouane rivers development.

There is controversy over the extent of the negative impact of these diversions on the environment, since the concept of minimum ecological flow remains controversial among biologists. Water does remain in Canada for each of these diversions, but this is not the point: Canadians collectively ignore the fact that their daily comfort and economic activity rest, to a certain extent, on major river diversions. It is therefore difficult to defend the argument that water transfers to the United States would weaken the environment, if one refuses to consider the phasing out of transfers within Canada.

At the end of 2005, with the signing of the annex 2001 implementing agreement, it was safe to say continental water diversions were definitely outdated as a water management concept. Demand was stabilizing in the western United States; public opinion had developed a strong distaste for the idea; governments were more than reluctant to invest the billions of dollars each scheme implied; diversion were by far not the most cost-effective way to provide for additional water volumes; and conservation and recycling were promising alternatives to the long-used supply management approach.

## Toward The Return of Massive Continental Water Diversion Projects

Climate change could shake this status quo. As years go by, and as scientists as well as local governments accumulate impressive data attesting to a definite trend toward climate change, the developing weather shift could prove more challenging than the long-advocated water exports to the United States of the 19605 and 19705. The possibility that climate change could affect Canada's water quality and quantity is serious enough that Environment Canada is spending several million dollars on scientific studies to try and assess the risk. Competing uses-for urban areas, agriculture, and oil production, for instance-in regions such as Alberta where water could become scarcer, could lead to conflicts between groups that would see authorities intervene with unpopular measures such as water pricing or regulation.

Present forecasts, so far proven by observed data, show increased global precipitation, but this extra amount of water would be more than offset by increased evapotranspiration (Et) because of higher temperatures. In the already dry Okanagan Valley in British Columbia, water demand for orchards could therefore be multiplied by 12 in the summertime because of increased Et. In addition, a change in the pattern of how precipitation falls throughout the year, coupled with changes in temperature, could very much alter its availability, resulting in seasonal but dramatic supply reduction, as happened in Québec in August 2002.

In the Prairies, already less snow falls in the winter, and it melts earlier in the spring, allowing for less water in rivers in late spring and summer. Along the eastern slopes of the Canadian Rocky Mountains, glacier cover has decreased rapidly in recent years, and total cover is now approaching the lowest experienced in the past 10,000 years. Rivers in the Prairies are mainly fed in the summer by melting glaciers. As the glacial cover has decreased, so have the downstream flow volumes. This finding appears to contradict projections of the intergovernmental panel on climate change (IPCC) that warmer temperatures will cause glacial contributions to downstream flow regimes to increase in the short term. Historical stream flow data indicate that this increased flow phase has already passed, and that the basins have entered a potentially long-term trend of declining flows. The continuation of this trend would exacerbate the water shortages that are already apparent across many areas of Alberta and Saskatchewan owing to drought. Climate change is of course not merely affecting Canada, but the United States as well. Similar conclusions have already been reached with regard to the potential impact of the Rockies' disappearing snow and glacial cover on the western United States. Planning for the impact of climate change remains difficult, because building accurate models for various regions-ones that give a true picture of future available runoff-remains a daunting task for programmers. What we now have are educated guesses sustained by general models and data trends. But these trends do show that several regions in the United States, especially in the dry west, could be faced with critical water shortages, given present demand.

This might prove to be a scenario in which the United States could become interested again in Canadian water: the goal would be to prevent a social crisis triggered by conflicts between agriculture and cities for dwindling water resources. The equation for the American federal government would be the perceived financial, social, and political costs of finding new water resources, as opposed to arbitrating against agricultural water needs. No one can predict what choice officials would then make if this scenario came true. The following parameters would be significant:

- Agriculture is responsible for 80 percent of water consumption in the American west.
- A 15 percent reduction in irrigation use would satisfy the needs forecast for cities and industries."
- Agriculture does not currently pay the market price for its water, thanks to strong federal subsidies; tariff increases would result in more disciplined behaviour and encourage conservation.
- Western farmers have begun selling or transferring their water rights to cities, as illustrated by the October 2003 water deal between the Imperial Irrigation District and the Metropolitan Water District in California.

Massive water diversions are not new in North America: they were extensively used in both Canada and the United States, in the latter mainly to develop the west and to bring water to cities and fields so as to sustain an imported way of life, and in the former to develop hydropower projects. Beginning in the 19605, western American states and engineering firms began developing continental water diversion projects because of fears that continued development would bring about severe water scarcity and social disruption. These doomsday scenarios did not materialize, due to reduced economic growthespecially in agriculture, as well as more efficient water use technology and the advent of water conservation, and the huge environmental and financial costs of these projects.

However, the very idea of massive water diversions is not completely dead. Tom Kierans, the designer of the Grand Canal, is still lobbying Ottawa and Washington. More challenging is the idea that ongoing climate change could dramatically affect agriculture in the American and Canadian west. Under these conditions, what answer would governments have? Whither agriculture? Would governments favour the more profitable and water-efficient farmers or bring back to the forefront the idea of transferring large amounts of water from humid parts of the continent to drier zones? The answer probably lies in the severity of the developing weather pattern change and in the adaptation capability of the agriculture sector.

## Domestic Water Supply in Rural Areas: Implications for Policy

Lack of access to safe water is at the heart of the poverty trap, especially for women and children, who suffer in terms of illness, drudgery in collection of water, and lost opportunities because of the time that water collection consumes. In rural Africa, according to the World Bank, 40 million hours are spent, each year in collecting water for domestic use and half of Africa's population is without access to safe water (Black, 1998). Recent renewed focus on poverty alleviation has resulted in increased attention to the benefits of improved water accessibility. Poverty assessment research has consistently shown that improvement in water services is a critical element in designing and implementing effective strategies for poverty alleviation. In Tanzania, experience with respect to rural water improvement has, in general, been an unmitigated disaster.

Water provision facilities have invariably fallen into disuse or disrepair because the approach used to run them failed to ensure sustainability of services. As a result, a new water sector vision has emerged based on the demand responsive approach (DRA). The major principle of the vision regards water as an economic and social good to be managed at the lowest appropriate level. The DRA recognises the inherent capacity of communities in taking greater responsibility for identifying and solving their water supply problems.

Adoption of the DRA by most water and sanitation programmes has changed the approach to evaluating their success and failure. Traditionally, water and sanitation programmes have depended on detailed blueprints to provide the basis for control and predictability. Benefit-cost analysis has been a major economic tool in evaluating these projects. However, the DRA broadens the scope of evaluation and blueprints cannot be drawn up since decisions are made jointly with communities, and problem solving is based on those partnerships (Narayan, 1995). Participatory evaluation becomes an essential tool in assisting attaining stated objectives and is an integral part of overall programme monitoring and evaluation activities. Participatory evaluation requires the utilisation of different tools and approaches in evaluating the sustainability of such projects.

This chapter is part of a follow-up participatory evaluation exercise of two community water supply programmes in Central Tanzania and involves presentation of the results on communities' willingness to pay (WTP) for improving the availability and quality of water services. WaterAid, a non-governmental organisation (NGO) based in the United Kingdom and the Lutheran World Federation through the Tanganyika Christian Refugee Service (TCRS), provide substantial support and expertise for such programmes. The former operates in the Dodoma Region while the latter operates in the Singida Region. The two regions constitute the semi-arid plateau of Central Tanzania.

It is anticipated that positive WTP indicates potential community ability to recover operation and management costs (Altaf et al., 1992). Thus WTP can be used to help ascertain the potential for fulfilling sustainability, at least from a financial viewpoint. A major objective of the study was to find out if communities studied were willing to pay an increased user fee or tariff for improving and expanding existing water services. If willing, how much (i.e. what percentage) could be added to the current user fee or tariff to achieve the required improvement? Based on the random utility hypothesis, a system of Multinomial Logit functions were developed and used to estimate individuals' willingness to pay and impact on revenue from increasing the water user fees or water tariff. The WTP was calculated by estimating the payments that would cause the respondent to be indifferent after a single unit increase in an independent variable. The estimated mean WTP and proportion of individuals who are able to pay were used to calculate the potential revenue that could be generated if the policy of increasing user fee or tariff was instituted.

In Tanzania, only 40% of the population has access to clean and safe water (MoW, 1997). Between 1965 and 1985, the Government of Tanzania (GoT) spent a substantial amount of money in developing rural-based water prefects to alleviate the water supply problem, but without notable success. Most of the investments ended in failure. Government, in collaboration with NGOs, is currently trying to revive many of these projects.

The government is providing technical assistance, while the NGOs supply initial capital requirements, overhead costs and supervision in their design and implementation. When the utility is operational, the beneficiaries (i.e. the villagers) either pay for the water from the utility or contribute an annual water fee, both of which are added to the village water fund. It is anticipated that the size of the village water fund will be sufficient to operate and expand the water utilities, based on market forces. To attain such sustainability-related goals, existence of efficient water markets is essential. Sufficient funds have to be generated to meet both overhead costs and investment requirements. Information on WTP for required improvements provides an indicator of the communities' potential capability of generating funds for such purposes.

Tanzania has sufficient surface and ground water to meet its present needs. Lack of capital and uneven availability of water across regions has limited Tanzania's capacity to utilise existing water resources to provide her population with clean and safe water. As population increases, conflicts in the allocation of water between energy, households, commercial livestock, and use in the agricultural sub-sector, increases. Weak water policies, inadequate community participation, and uncoordinated donor support has resulted in wastage or inefficiencies in the form of failed water development programmes.

After independence in 1960, donors supported large-and small-scale water development schemes. However, communities participated little in their planning and management, and also contributed little to both capital and operational/management costs. The policy of the government at independence was 'free, clean and safe water for all,' the objective being to provide clean and safe water to all villages in rural Tanzania by the year 2000 (MoW, 1997). As indicated above, this policy has now been abandoned in favour of a demand-driven water development programme. The new water policy enacted in 1997 (MoW, 1997), promotes the provision of efficient, affordable and sustainable water supplies and sanitation.

The emphasis is on community planning and management, private provision of goods, works and services relating to water supply, and public sector regulation, facilitation and environmental management. Water supply or water development programmes have to be demand-driven, and the community must be willing to participate in decentralised management and cost sharing. The Ministry of Water (MoW) has been restructured and its role has evolved from that of a water service provider to that of a regulator and a facilitator.

The new water policy underscores the importance of community participation and management in ongoing and new projects. In order to improve water supply services at the village level, the roles of MoW and other stakeholders can be summarised as follows (MoW, 1997):

- Village level. Small-scale water supply projects are to he operated and managed at the village level. Operation and management costs are to be met with funds raised within communities. The emphasis is on management by participation through formulation of village water committees that oversee and manage the utilities on behalf of community members. The communities agree upon the operational modes, with specific emphasis being paid to women's participation at all stages of water project development and management.
- District level. The water department at the district level is required to facilitate training of water managers and attendants/mechanics at the village level. The department is expected to maintain a pool of experienced technicians who will collaborate with village level mechanics in servicing and repairing established utilities. Another role of the water department at the district level is to make sure that necessary spare parts are available when needed.
- Regional level. The water department at the regional level has jurisdiction in providing guidance and making sure that government water policy and rules are adhered to. Apart from managing water supplies at the regional headquarter level, the department has to provide consulting services to districts

and villages in terms of training, and facilitating availability and distribution of spare parts.

- National level. At the national level, the MoW has responsibility for financing and managing large-scale water supply programmes (especially in large cities), to train water professionals, and to finance maintenance units at the district and regional levels. Other responsibilities are to standardise capital equipment and tools used in water development, to help facilitate availability of spare parts, and to coordinate donor and NGOs activities to make sure they follow Tanzanian water policies and local government rules and regulations.

The general objective of the water policy is to propagate the new vision of community participation and management based on the DRA. Communities arc expected to appreciate that water supply is no longer a free service provided by government, but rather government or other development agencies are available to facilitate and complement community efforts at meeting their own water supply needs. Thus, this vision is currently guiding the development of community-based water supply projects at the village level in the Dodoma and Singida Regions of Central Tanzania. Detailed descriptions and the rules pertaining to the programmes in each region can be found in Kaliba (2002) or the WaterAid website.

In both regions, financing for the water projects is thus divided among the beneficiaries, the government and donors. Villagers make cash contributions towards capital costs. They contribute time and labour, local materials and contribute to hospitality costs of visiting government staff. The initial amount of money contributed to a project by a village has been standardised and agreed by the major donors working in the regions.

The cash contribution is not currently expendable but acts as a reserve fund or safety valve for the project. The community-based water fund has to be increased using profit margins generated from user fees (Dodoma Region) or tariffs (Singida Region) when the project is operational. The government provides professional staff, an annual cash contribution towards the programme, and provides some transport and most of the construction equipment. Donor (WaterAid in the Dodoma Region and TCRS in the Singida Region) funding pays for the purchase of locally procured materials, government staff fieldwork allowances, training courses and running costs. The donors also procure all imported materials, equipment and vehicles, and employ technical and management back-up staff. In the Dodoma Region the village water utilities have emphasised deep boreholes and engine-driven pumps, and charge a tariff (i.e. a specific amount per litre of water used). In contrast, in the Singida Region, much greater emphasis is placed on shallow wells (8-10 m deep) and a user fee (i.e. a specific amount per year per household), is charged. In the Dodoma Region, between 1991 and 1999, 857 projects were implemented under the WAMMA1 project funded by WaterAid. During this study period, 29% of the projects were not functioning. For those not working, 75% were waiting to be repaired, and 25% had permanent problems (e.g. missing engines, damaged pumps or dried up waterholes).

Useful indicators of progress are the growth in the populations served and village water funds voluntarily contributed by communities. The total population served increased from 0.27 million 1994 to 1.23 million in 1999. This is an increase of about 320% over a period of six years. During the same period, the contributions to the village water funds increased, in real Tsh terms, from about 0.24 million in 1994 to about 14.87 million in 1999. Since there is no evidence of a regional increase in purchasing power, and given the low income per capita in the region (BoT, 2000), this is a good indicator of the success of community-based water development projects in the region.

During that period, TCRS constructed 49 boreholes, 465 shallow wells, and three windmills. The number of villages with water committees increased from three in 1986 to 244 in 1995. In the same period, three villages did set up water committees without the help of TCRS. The number of villages with water funds increased from zero in 1986 to 229 in 1995. Although these numbers indicate the magnitude of time and money invested in water utility projects, they of course, do not provide by themselves any explicit evidence of the potential sustainability of the projects and/or progress towards achievement of sustainability goals.

To elicit WTP information from individuals, the contingentvaluation (CV) method is used. There are three major procedures used in CV surveys, namely conjoint analysis, dichotomous choice and the payment scale approaches. The conjoint analysis procedure asks respondents to rate rather than to price alternatives. It allows for measurement of consumer preference between items with multiple attributes (Baidu-Forson et al., 1997). In the dichotomous choice approach, respondents are asked whether they would vote to change the provision of some public good at a cost of $X to themselves.

The respondents answer yes or no. For the respondent who answers yes, the experiment is repeated by varying X, until the respondent says no. After that, the respondent is asked to state the value he/she is willing to pay for the good or service if provided. For the respondent, the amount mentioned is considered the maximum WTP for that good or service. The payment scale procedure allows respondents to choose a value or price (cost) on a given scale measure. A range of values are presented to the respondents and they are asked to identify one value or price (cost) they are willing to pay or incur to purchase the good or service being evaluated. The respondent's choice from the given scale indicates the WTP for the stated good or service. Positive values indicate positive demand, and zero values indicate zero demand for the good or service.

In all cases, the surplus benefits are calculated using the parameters estimated from a model based on the type of distribution attached to WTP. In most cases, the models and type of distribution to use are dictated by the type of questions imposed and responses obtained. Other good references are those of Rosenberger and Walsh (1997) for ordinary least squares models; Kenkel and Norris (1995) for models with log-normal distribution; Steven et al. (1997) for Tobit models; and a classical work by Haneman (1984) for Probit and Logit models. CV surveys entail three operations: designing a questionnaire, conducting a survey, and analysing the results. The validity of the estimates depends on how skilfully these operations have been done (Christe and Schwab, 1995). Donaldson et al. (1997) suggest that the use of a payment scale is more valid than the open-ended or closed-ended approach when presenting hypothetical bids. The limitation of the payment scale is constructing scaling benchmarks. This limitation can be reduced through using available market and nonmarket information to construct the scales, as suggested by Donaldson et al. (1997).

From the above review, it can be seen that there are different approaches to modelling WTP. Nevertheless, all procedures involve the respondent choosing one option from a range of other alternative services or goods based on their expectations. In order to develop models that are consistent with economic theory, the McFadden (1981) random utility hypothesis has become a standard approach to modelling WTP. As far as our study was concerned, the objective was to find out if communities are willing to pay an increased fee or tariff in order to get additional money for improvement or expansion of existing water utilities. If willing, how much (i.e. what percentage) should be added to the current user fee or tariff to achieve the required improvement? Field data were collected from 30 villages that participated in community water utility projects for a period of three years in Dodoma and Singida Regions (15 villages in each region) prior to the field surveys. Dodoma Region was divided into five clusters based on tribal distribution. From each cluster, three villages were randomly selected, differentiated according to accessibility. The results of past participatory rural appraisals (PRAs) conducted by WAMMA (some as early as the 1999) were reviewed and used in making adjustments to the questionnaire and in helping to develop pre-coded responses and socioeconomic indicators. In the Singida Region, due to transportation problems (part of the El-Nino legacy), only Singida Rural District was included. Rased on information gathered at the District and Regional Water Departments and the Regional Development Office, the district was divided into three agroecological zones. From each zone, five villages were identified to form a sample cluster. The Singida Regional Development Officer also had some PRA results that were used to fine-tune the questionnaire used in the study.

Two survey instruments were developed to gather the necessary information — the village checklist and the structured questionnaire. The village checklist solicited general information at the village level. This was done through meetings and group discussions attended by community members, government leaders, project managers, pump operators, extension agents, and water committee members. The major objective of the checklist was to get a general overview of the village projects based on group discussions. The discussion also helped to pre-test the structured questionnaire and in further adjusting the precoded responses. The questionnaire was administered to 225 individual respondents in each region. In each village, 15 respondents (at least five men and five women) were interviewed. The respondents were selected randomly from the village register and grouped based on age and social status of the individuals. Sample selection was stopped when there were enough sample respondents in each group. The two survey instruments and further information about the approach used in the study can be found in Kaliba (2002).

Regarding information on participation and management, several questions were asked. The questions related to the source of the project initiative, personal involvement in decisionmaking, labour and financial contributions, and the involvement of women in project design and implementation, emphasising participation in decision-making and informed choice. Questions were also asked regarding consumer satisfaction with the services and management of the water utility. The questions focused on: getting information on personal involvement in management; rating of the water managers' performance; knowledge of how the water management structure functioned; and the possibility of influencing change. To quantify respondent participation in project activities, and satisfaction in project performance, individual's responses were aggregated and used to calculate participation and satisfaction indexes as suggested by Sara and Katz (1998) and by the World Bank/UNDP (2000). Once again, further details on the approach used as far as this chapter is concerned can be found in Kaliba (2002).

On the WTP section, a respondent was asked if there was a need to improve the current water service. The answer was either 'yes' or 'no'. If yes, what kind of improvement was needed? The respondent was encouraged to identify only one improvement. After the respondent identified the type or choice of improvement required, she/he was asked to suggest an increase in tariff (for the Dodoma Region) or user fees (for the Singida Region) in order to cover the costs of the proposed improvement.

As explained above, the analysis was directed at identifying types of water services that need improvement and ascertaining the WTP for the improvement. In the Dodoma Region, the results concerning the water utility service can be summarized as follows: about 14% of respondents indicated that they were satisfied with the status quo (no improvement); and 81% suggested increasing pump capacity or buying a more powerful engine (thereby increasing the rate of water discharge). Thirty-three percent of respondents proposed increasing the number of water distribution points (water reticulation), and 22% proposed other improvements such as fencing, construction of new water tanks, construction of security houses, and construction of livestock watering points. The suggested tariff increase ranged from 5 to 100 Tsh/20 1. The existing tariffs averaged 20 Tsh, with a standard deviation of 9 Tsh (per 20 1 bucket of water) and on average amounted to a total of 895 Tsh/household/year.

In the Singida Region, 31% of the respondents were satisfied with the status quo, 80% wanted deeper wells that can produce more water, 29% wanted to increase the number of hand pumps in their village for water distribution purposes, and 10% indicated other types of improvement enumerated above. The suggested increase in a user fee per household per year ranged from 20 to 200 Tsh. The average user fee was 503 Tsh per year with a standard deviation of 170 Tsh. During the analysis, variables for Clusters 4 and 5 in the Dodoma Region and Cluster 3 in the Singida Region were not included in the models due to correlation problems. The likelihood ratio statistics that test for no effects was statistically significant at the 5% level in both the Dodoma and Singida Regions models.

The positive and statistically significant variables in the Dodoma Region were family size and satisfaction in the performance of project activities. These results imply a respondent residing in a relatively large family or who was satisfied in the project's performance were more likely to answer 'yes' to the question regarding the need for water service improvement. A large family implies the need for more frequent water collection trips. Improvements relating to reducing congestion at watering points (increased pump capacity or number of watering points) are likely to reduce time and effort expended in water collection. Satisfaction with reference to project performance implied both an increased demand and willingness to commit resources for improvement.

Negative and statistically significant variables were age, wealth and cash contributions. Older people and richer respondents were thus more likely to choose to maintain the status quo. In the study area, older people are less likely to be directly involved in water collection activities, while rich respondents could conceivably have access to other water sources (e.g. have their own wells) or delegate others to collect water for them. The negative sign on the cash contribution variable can possibly be attributed to 'contribution fatigue.' Respondents who contributed more during project initiation or development were more likely to say 'no' to the improvement question.

Finally, respondents in Clusters 4 and 5 were more likely to be predisposed to commit resources to improvements, because the general performance of the projects in these clusters was relatively poor (Kaliba, 2002). The underlying implications of the above results for water utility projects in the Dodoma Region are that projects whose performance creates a sense of satisfaction among consumers and a strategic water fund that is rising (so as not to create contribution fatigue) are critically important ingredients in generating community-related resources for water-related improvements. In the Singida

Region, females were more willing to pay for improvement than male respondents, not surprising since they are primarily responsible for water fetching activities. The fact that this variable was not statistically significant in the Dodoma Region was unexpected. However, in the Dodoma Region water utility projects produced a more reliable water supply and there were generally fewer complaints about their performance (Kaliba, 2002). Another important and positively significant variable was the level of satisfaction concerning the performance of the project.

The Multinomial Logit model was used to model factors affecting choice of services: maintain the status quo, increase rates of water discharge (pump capacity or depth of water well or more powerful engines), increase the number of water distribution points/water reticulation, and other improvements, as explained above. In constructing the Multinomial Logit models, the parameters of status quo variables were set to zero. Again, the signs on the estimated parameters show the direction of the marginal effects on choice of improvement desired. The comparison is related to the deleted choice. As an example, for Dodoma Region, respondents who perceived that water service quality was high were more likely to say 'no' for an increase in water supply or the number of water distribution points but were more likely to say 'yes' for other improvements. The interaction between quality and proposed bids were important in making the choices or indicating the type of improvement desired.

The opposite was true for the Singida Region where the interaction between quality and bids influenced the choices negatively. There, the important factor that significantly influenced the choice of improvements was individual bids. This may be related to the technology and poorer level of services in the Singida Region. Respondents may not be willing to spend more money on failing projects. Singida Region projects will have to increase performance in terms of quantity and quality of water supplied, in order to motivate community members to contribute more resources toward project development and expansion. This was particularly important in some villages where community participation in project initiation development was low (Kaliba, 2002).

Again, the reference is the status quo bid. The value of-0.145 in the quality variable in Cluster 1 of the Dodoma Region implies an increase in the quality of water by one unit decreases the percentage of respondents who want increase in water supply by 14.5%. About

64% of respondents in Dodoma and 59% of respondents in Singida voted for an increase in water supply through increased pump capacity or number of water distribution points (water reticulation). Most villages have only one watering point and it is clear that a lot of time and effort is required for water-collection activities.

### Estimated Mean Willingness to Pay

Respondents who wanted to increase water supply in Dodoma Region were willing to pay 82 Tsh per 20l of water above the existing tariff. Respondents whose choice was improving water reticulation possibilities and other types of improvement were willing to pay 29 and 5 Tsh above the existing tariffs. In the Singida Region, the analogous amounts were 91, 68 and 80 Tsh per household per year, respectively. A joint t-test failed to reject a null hypothesis that there is no gender difference on willingness to pay in both regions at the 5% level of significance. In the study area, water fetching is not a man's responsibility. However, they care. Time wasted by women and children in the search for water has a negative impact on their other activities and therefore has a compounding effect on all members of the household.

### Potential Revenue from Communities

Potential revenue that can be generated for project improvements was estimated using the estimated mean willingness to pay, proportion of respondents who chose the services, and reported average daily consumption of water per household. It can be seen that there is the potential for generating much more in the way of community-derived funds in the Dodoma Region compared with the Singida Region. For the Dodoma Region, the amount that could be generated is substantial by Tanzanian standards.

In fact, the amount may be more than the budget allocated by the central government to regional development programmes in all sectors of the economy. However, a note of caution is in order. Care will be needed to implement policies aimed at increasing revenue generation for water improvement purposes. Higher tariff or user fees may exacerbate inequitability of access to water services because poorer households will undoubtedly be more price-sensitive than richer households. Although there are potentially mechanisms to subsidise the poor and the elderly, such arrangements tend to alienate those affected and limit their willingness to participate and be involved in decision-making processes.

## Policy Implications

Results for the WTP analysis and on proposed choices for improvement, imply that respondents were eager to improve the ease and efficiency of the water collection process. For most respondents in both regions this was explicitly articulated in the form of a desire to increase water supply and the number of water distribution points. Most villages have only one or two watering points, and it was clear that a great deal of time and effort is expended in water collection activities.

There is some potential for increasing project revenues to address such needs through increases in water tariff or water user fees. However, due to the differences in the technologies to access water, it appears that substantially more funds can potentially be generated for such purposes in Dodoma than in Singida. The water technology being used in the Singida Region is apparently not solving the problem of the dry season water scarcity, let alone the issue of water quality. A new approach and vision is needed to provide a long-lasting solution to the problem in that region. Dependency on shallow wells seems to have less potential for contributing to financial sustainability of water ulility projects in the Singida Region than the deep boreholes in the Dodoma Region. Whatever happens, in both regions, if increases in water utility levies are planned in order to improve water services, two issues will need to he carefully considered. One is to take into account the views of women especially with respect to the location of new watering points or wells. The other is to ensure that any adjustment in water levies does not create undue hardships for the community or unduly inhibit access by the more disadvantage families/individuals in the communities (i.e. the poor or elderly).

In conclusion, project sustainability, at least from a financial viewpoint, will be largely determined by the degree to which it continues to deliver its intended benefits over a long period of time. As far as Central Tanzania is concerned, there is some potential for developing sustainable water supply programmes based, to a large extent, on a community cost recovery mechanism. In villages where there has been strong satisfaction with respect to water project performance, this chapter has demonstrated that individuals are willing to contribute more resources for improvement. It is important to note that, in these villages, community members are highly motivated and participate strongly in the daily management of the projects. An increase in community participation and changing of the water provision technology for the Singida Region is important for future project expansion and new project development.

## Role of Virtual Water in Solving Water Scarcity

Water, as a source of food security, plays an essential role in ensuring sustainable food resources for a growing population. However, water scarcity is a common problem in many countries and regions. Because of the uneven distribution of water resources and population densities worldwide, water demand already exceeds water supply in nearly 80 countries that support more than 40% of the world's population (Bennett 2000). Water scarcity has become an increasing constraint to their economic development, particularly food production-the biggest water user (Hong et al. 2002). Nonrenewable fossil water has been exploited to relieve the immediate pressure of water stress; this use has caused the depletion of resources and obstruction of future economic development and food security. To combat water scarcity and increasing food production, many countries have made efforts to expand irrigation.

Despite a tremendous effort, the coverage ratio of irrigation in most of these countries remains low, simply because of their scarce water resources. The growth of food production overall has lagged and has been unstable (FAO 2001). In contrast to food production, the growth of population and food demand has been rapid in China. Meeting the water-food challenge is difficult in China because of the increasing population and water demand. In order to relieve water stress, many investment projects in China are planned and have been carried out, in order to focus on changing the uneven distribution of water resources, and to fill the gap between water supply and demand.

The role of virtual water through food trade as a solution to water scarcity is not recognised in China. Indeed, most researchers do not associate the international food trade with food security. However, they now advocate regional food trading as a solution to water stress in Northern China. These opinions are consistent with Chinese government policies of food security. The considerations of the Chinese government for food trade include domestic food security and social stability, the impact on rural employment when grain production remains the major provider of employment in rural areas, and the pressure of rural-urban migration and the social economic impacts on cities (Hong 2002).

## Water Deficit and Demand

### *Water Scarcity in Agricultural Production*

The increasing demand for food needs two basic natural resources for its production, arable land and fresh water. Future food production depends on the availability of adequate and sustainable water resources. Agriculture is responsible for about 70% of all freshwater withdrawn, and it is expected that more water will be used for irrigation as world food production increases.

An FAO study of 93 developing countries indicates that some water-scarce nations are already withdrawing water supplies faster than they can be renewed (FAO 2004). Over the years, China has maintained a population increase of 1.4% and an increase in economic development of 9% per year through using a great deal of water with low efficiency. However, agricultural irrigation is constrained by water scarcity and pollution, with 70% of food output produced on irrigated land, and 73% of the total available water used for irrigation. Furthermore, agriculture is in fierce competition with industry, municipal, living and environment water uses.

Actually, in China, agriculture competition for water is below industrial and domestic demands (Hong 1999) because the direct economic output of agricultural production is rather low. In China, every $10^4$ $m^3$ of water feeds 4.46 persons.

North China, Inner Mongolia, Hebei, Beijing, Tianjin, Shanxi, Henan and Shandong, seven provinces, use about $168.5 \times 10^9$ $m^3$, which only can feed $75 \times 10^6$ persons, i.e. 23% of the total population. The demand for food by the population is met by food produced in Southern China, where farmland is scarce although water resources are relatively sufficient. Present resources meet the demand for food by the large population. Indeed, the requirement for freshwater resources is four times higher than what is normally available. The required capacity for freshwater resources is the lowest of all the resources estimated (Qu 1992).

At the national level, it is estimated that there may be enough water available, but some regions will face serious water shortages, with inefficient use and serious pollution. In northern areas, water scarcity and water pollution are serious while floods and water pollution are the main problems in southern areas. 95% of available water resources in the north have been exploited, while 6% of water resources have been used in the south. Water consumption in agriculture increased in North China where land is not a limiting factor for food production. Because water use efficiency is on average 0.4-0.5, irrigation water in the north is $7.5 \times 10^3$-$12 \times 10^3$ $m^3$/ha, or 2-5 times higher than actual crop water demand (Liu 2000). The management of water use is expected to improve for avoiding water waste and enhancing irrigation capacity.

Water scarcity in the north has resulted in higher opportunity cost than that in the south. The huge investment in the South-North Water Transfer Project also supports this viewpoint. For water-scarce China, water use efficiency is much lower than that of developed countries. With the rapid population growth, water stress is becoming a formidable challenge. Meanwhile, when water resource load capacity reaches the limit, direct cost of enlargement of food production will quickly increase and result in native food price rises. The most plausible result is that the cost of water for agriculture becomes much higher because of competition between sectors.

### *Future Water Demand of Agriculture*

Agricultural water consumption is the principal source of total water consumption, and is increasing with growing food demand. Irrigated area has increased, but total irrigation water quantity and withdrawal do not balance. This is due to irrigation efficiency enhancement and new irrigation technology improvements. Total water consumption has increased with unchanged available agricultural water quantity. The increasing water demands of economic development and domestic needs make additional water for agriculture more difficult to find, especially in North China. The large acreage of crop production in this area indicates a large irrigation requirement.

Because different crops have different irrigation water demands and different arable land acreages in North China (Qian et al. 2001), total irrigation water demand is $48.55 \times 10^9 m^3$; $5.00 \times 10^9$ $m^3$ for rice, $26.53 \times 10^9$ $m^3$ for wheat and $17.02 \times 10^9$ $m^3$ for maize. Actual irrigation efficiency in agriculture is quite low (0.4-0.5), thus, actual water demand required to support production in North China is about $97.1 \times 10^9$ tonnes. The ratio for agriculture water use accounts for about 57.63% of total water resources in North China, which irrigated only 53.12% of its total ploughed acreage. The remainder of the ploughed acreage irrigation demand, as well as other sectors' demands, intensifies water stress, influencing economic development and agricultural production. A water transfer project was selected to alleviate water pressure in North China, but in the view of the economic cost, virtual water trade through food trade may be a better solution for water deficits.

## Food Demand

### *Population Growth*

The population in China is about 1.3 billion and has been increasing. Expansion of population influences every aspect of social development. Insufficient natural (fresh water, land, forest, etc.) and productive resources (e.g. food) originate from population growth. Finite resources cannot support the huge demand of the growing population. Resources allocation ratios are very small. These problems interlace and restrict economic development and social advancement. At the same time, potential risks are concealed in the problems. The population control policy, which confines the number of children in a family, cannot restrict overall growth in the short term because the growth is based on an existing huge population. It is predicted that population growth will be negative in 2050 (Kang 1998).

The most important problem is that China must have enough food for the population, whether from food production or imports. Freshwater resource scarcity and uneven distribution strain food production, resulting in an increasing portion of domestic food supply coming from imports. Such imports, in essence, have substituted for the water that would otherwise be needed for producing the food locally. Studies of virtual water and food security in Mediterranean and Africa countries have shown that water scarcity in these countries resulted in food insecurity. China has the same food issues, and uneven distribution of water and land resources has resulted in food supply scarcity. Furthermore, there is rising competition between water needed for agriculture and water for domestic use in burgeoning cities and ever-expanding industrial developments. Consequently, today the world must act wisely to conserve, preserve and improve its water supplies (FAO 2004).

Food demand is a function of the size of population and consumption per capita. Rapid population growth will inevitably drive up future food demand, while per capita food consumption is likely to increase as income rises, adding an extra demand for food (Hong et al. 2002). The growth in population and foodstuff demand per capita has jointly promoted the growth of total foodstuff demand in China

(Kang 1998). Foodstuff demand requires that food supply increases concomitantly. However, expansion of food production is limited by farmland and water resources; thus, the pathways of food supply enlargement are expected to change. Increasing output and food import is required to solve food scarcity, together with water resources, to meet the food demand of the population. Water resources and food trade are complex issues for society.

## Crop Production and Food Consumption

Cereal production in China provides the basic energy requirement directly as food and indirectly as feed for animals that are used for human consumption. The increase of grain production, including rice, wheat and maize, has been rapid since 1980. The trend in grain production up to the end of the 1990s was smooth, with only small increases from 1991 to 2000 because land productivity is approaching the limit, and no new productive technology has been created to promote increasing food production. The production of cereals, cotton, oil and sugar was either unchanged or declined, while the production of fisheries and livestock products has rapidly increased.

Before 1997 most cereal production was consumed domestically, typically rice, since then, the diet has changed and more non-cereal products are being consumed. With increasing income, meat, egg and milk demand has increased while cereal consumption has decreased. In contrast to decreasing direct cereal consumption, meat consumption has resulted in indirect cereal consumption by livestock. At present, according to the current GDP per capita, China is positioned by the World Bank in the low-middle income category. However, because of rapid economic development, per capita dietary structure indicates that life quality change is one of the most important variables that influences consumption.

### *Food Trade in the Past Two Decades*

Food imports in China have fluctuated, especially wheat imports. This import demand is decided by domestic production capacity, and is a way to ensure food supply. Production technologies, especially water saving technology and multiple crop index improvement, have promoted the productivity of wheat in China. However, the increase in food production is still limited by water and land resources. Although irrigation efficiency has generally improved, insufficient water resources are allocated to agriculture. Food products imported into China from 1980 to 2000 were mainly in the form of wheat and maize.

Net import volume has decreased gradually for wheat from 11.59 × $10^6$ tonnes in 1995 to 0.88 × $10^6$ tonnes in 2000, and for maize from 5.18 × $10^6$ tonnes in 1995 to zero in 2000, because domestic supply has increased. However, in contrast to declining imports, the volume of export products has increased from 1980 to 2000 due to increasing production because of improved agricultural technologies. Growing export products include rice, cotton, pork and fishery products, in which rice export rose from 0.05 × $10^6$ tonnes in 1995 to 3.75 × $10^6$ tonnes in 1998. Pork exports declined from 3.24 × $10^6$ tonnes in 1982 to 1.96 × $10^6$ tonnes in 1999 because of increased domestic demand.

With the limited potential for additional water supply, the pursuit of water sources for the improvement of water use efficiency has shifted to high-value crops. Vegetables and fruits are typically encouraged in China as economic agricultural products and as a way of raising rural incomes. The exports of vegetables and fruit increased from 1980-2000. This shift has produced high economic returns and made possible more selections of imports for the alleviation of the intensification of water stress.

## Analysis of the Role of Virtual Water

Allan defined 'virtual water' as the water embodied in food crops that are traded internationally. Allan (1997) termed such food imports as 'virtual imports' due to the fact that they are equivalent to a transfer of water to the importing countries. The concept is used to describe a water-scarce nation's opportunities for achieving food security by purchasing a portion of its food requirements in the international market. It compares the amount of water embodied in a crop that can be purchased internationally with the amount of water which would be required to produce that crop domestically. Virtual water is a term that links water, food and trade (Allan 2003). As the metaphor of virtual water gained currency in the mid-1990s, economists at the World Bank started to use another term the 'water, food and trade nexus' (McCalla 1997).

This combines agronomic, economic and political outlooks, with emphasis on water as the key factor of production. The agronomic component involves the opportunity cost of water, which may include production of alternative crops or use in municipal, industrial or recreational activities. The opportunity cost of water use, which is a key component of the 'virtual water' perspective, must be considered when seeking an efficient allocation of scarce water resources. The same concept applies to the water embodied in other productive activities, such as manufacturing and power generation. Some manufactured goods require a greater amount of water in the production process than other goods. Nations in which water is particularly scarce may gain by importing water-intensive manufactured goods, while exporting goods that require less water in production. Similarly, nations with limited freshwater resources may gain by generating electricity in coastal areas, using ocean water for cooling, or by importing electricity from countries with abundant hydroelectric resources.

### *Policy Relevance of Virtual Water*

Food trade plays an important role in redistributing the water resources of the world. The role of food imports in compensating water scarcity is elaborated with a quantitative accounting of the volume of virtual water embedded in their food imports. It is suggested that water be considered an economic good (Hoekstra et al. 2003a,b), and virtual water trade be seen as an instrument to achieve food security and efficient water use (Hofwegen et ai, 2003). However, when water policy issues are discussed, further implications of virtual water can be considered. Food imports are used when policy makers care about food security, while the feasibility of food imports is judged by opportunity cost. Wichelns (2003), however, thought that the optimal trading strategy could be determined by examining the comparative advantage, not the opportunity cost. This comparative advantage includes consideration of the water resource endowment and the production technologies in two or more countries that can engage in trade.

On the other side, virtual water has geopolitical implications. Trade induces dependence between countries, and it is influenced by and has implications on tariffs. Indeed, the issue of virtual water is related to globalization, which raises concerns among many politicians and the general public. Increasing global trade implies increased interdependence of nations (Hofwegen et al. 2003). Geopolitical conditions in food trade are not ignored because food security, like water security, is a highly contentious issue that needs carefully examining in terms of the degree and method of food self-sufficiency. Domestic political power has decided traded objectives, volumes and formations (Rang 2000). Bouwer (2000) suggested that an OFEC (Organization of Food Exporting Countries) be established with international controls, and in which the importing countries are ensured that global distribution of food will not be used as a political weapon. This study emphasizes the potential role of virtual water trade on the solution to water resource scarcity and food security.

### *Virtual Water Equivalents in Traded food*

Food production is constrained by water scarcity; thus, importing food compensates for scarce water resources. Virtual land is also evaluated using imports divided by yields per hectare in the same year from the FAO statistics, where virtual water is the amount of imports multiplied by the global average of virtual water in every crop (Hoekstra and Hung 2003). On average, $8.9 \times 10^6$ tonnes of wheat and $0.9 \times 10^6$ tonnes of maize were imported annually between 1980 and 2000. An average of $10.20 \times 10^9$ $m^3$ water and $27.7 \times 10^6$ ha of land would be required to produce this imported wheat, while an average of $0.32 \times 10^9$ $m^3$ water and $1.6 \times 10^6$ ha land would be required to produce the imported maize. The sum of $10.52 \times 10^9 m^3$ of virtual water is equivalent to 23% of the annual average transfer of water volume for the South-North Water Transfer (SNWT) project. From 1995 to 1999, net virtual water import volume of crops in China was calculated at $101.9 \times 10$m, and China became the fifth in the top-30 table of net virtual water countries (Hoekstra and Hung 2003). The sum of $29.3 \times 10^6$ $m^3$ of virtual land is equivalent to 19% of crop arable land in 2000.

### *Water Compensation Through Virtual Water*

Water resource scarcity constrains food production, and food supply cannot completely depend on the enlargement of food production. The basic food supply comes from food production and imports. Food import is appropriate when food price is lower than domestic production cost in the international market (Allan 1999). For virtual water embedded in the cereal imports, an aggregate ratio of 1 kg grain to 1 m of water is used. As such, the figures for cereal imports can be transformed into the same numeric volumes of virtual water in m/ per capita. Research shows that virtual water volume embodied in cereal is higher than the global average (Oki et al. 2003). This means that imports of virtual water in China should be greater than estimates in this study. The factors that influence opportunity cost of production caused by the growth of irrigation water volume to enlarge the food supply have two aspects. One is the food supply capacity including food demand, food alternatives, food productive capacity and native import capacity. The native import capacity includes international food prices, native foreign trade conditions and international aid. The other aspect is concerned with water resources, embodied in water use efficiency in agriculture, agriculture management, water resource volume, water pollution, wastewater treatment and reuse. Those limiting factors are complex and, when combined, the opportunity cost in different nations and regions are obviously different. The amount of agricultural produce derived from a certain amount of water is subject to a large number of variables. These include the geographic location of the region where crops are grown, the irrigation system used, the management of the irrigation system, soil types and climatic conditions (Turton et al. 2002).

Moving virtual water will be much cheaper than moving water itself (Bouwer 2003). Global movement of food from water-rich to water-poor countries should be just as feasible as moving petroleum products from oil-rich to oil-poor countries. Virtual water opportunities for saving local water resources include imports of electric power from areas with more abundant water for cooling of thermal power plants, or with dams for hydroelectric power production (Bouwer 2000). Currently, most researchers in China think food demand and supply in China are on firm ground. Some agreed that the food import dependence ratio might be augmented to 20% (Wang et al. 2002), while others argued that food supply should be self-dependent to ensure food security and food trade between regions within China (Cheng 2003).

Virtual water is useful to alleviate water scarcity pressure, but many governments are worried that over-dependence on other nations will result in a food crisis and social instability. Importing virtual water embedded in traded food in China can be seen to alleviate water stress, and even achieve food security. Virtual water trade may compensate for water demands of not only the past but also the future. Meanwhile, the trade can store water in its virtual water form, especially by food storage, to play a potential role in solving food problems. However, policy considerations of domestic water and food security are needed to promote better understanding of the measures that need to be taken to ensure their future supply.

# Chapter 8

## Managing Water Policy

Until recently, the eastern United States has been blessed with an abundance of water; unlike the arid West, shortages in the East have historically been "rare and short-lived." During the past few decades, however, water has increasingly become scarce, due to recurring droughts and burgeoning urban and suburban populations. A severe drought struck the East Coast in 2002, forcing residents in rural Maine to stop flushing toilets and washing dishes, significantly damaging crops in the Midwest, and leading New Jersey to ban lawn watering. But that drought was only the latest and harshest of several that has hit the region since the 1980s. Rapid population growth and commercial development in and around eastern cities have also contributed to shortages, in some cases leading to interstate disputes.

Scientists predict that erratic precipitation patterns will persist, producing recurring droughts in years to come; continued growth in demand will only exacerbate the consequences of weather shifts. For policymakers, water shortages present three central challenges. First and most obviously, scarcity reminds us of the limits of our natural resources, and forces us to consider conservation measures. Second, scarcity requires that we assess the efficiency of water allocation and usage. Third—especially in the context of water, a universally necessary resource—equity demands that basic needs be met, and that one group of users not be allowed to exclude another group from an essential resource. In recent decades, water shortages have compelled eastern lawmakers to face these three issues—conservation, efficiency, and equity—directly. The result has been growing dissatisfaction with the legal regime that had historically governed the distribution of water rights in the East: common law riparianism. Under riparianism, "an owner of land abutting a waterbody has the right to have the water continue to flow... subject to the equal rights of each owner to make strictly limited use of the water." Riparian landowners, also called riparians, "share the common benefits that arise from adjacency to defined bodies of water."

Although riparian law had governed surface water in the East from the time of the first Anglo-American settlers, it came under increasing criticism during the latter half of the twentieth century. From a conservation standpoint, riparianism failed to protect against excessive diversions by riparians, or to ensure minimum stream flows for the public. From an efficiency standpoint, riparian rights were inherently uncertain because they were correlatively defined, and thus could shift over time as neighbouring users and uses changed; uncertainty, in turn, inhibited investment and prevented the development of markets in transferable water rights. And from an equity standpoint, riparianism was accused of favoring private agrarian interests.

Moreover, because rights were determined by ex post litigation, riparianism—like any common law regime—lacked a procedural mechanism for reallocating resources to those in need during times of scarcity. In the hopes of addressing these doctrinal shortcomings and better dealing with water shortages, approximately half of the eastern states formerly governed by common law riparianism established water permit systems, also known as permitting systems. These regulated riparian statutes—often adopted in response to scarcity concerns —envision a rationalized, administrative approach to water rights allocation. In theory, regulated riparianism requires that all direct users obtain a limited-duration, renewable permit from a state administrative agency.

This presumably allows the agency to make comprehensive ex ante decisions about preferred uses and impacts on other users, rather than resolve single disputes in ex post litigation. Agencies also have authority to reallocate water during times of shortage. While these regulatory schemes are usually meant to be comprehensive, certain groups of users are often exempt from the permit requirement. In contrast with common law riparianism, use on nonriparian land is generally allowed under permitting systems. What has become apparent over the last few decades, however, is that regulated riparianism suffers from its own weaknesses and, more importantly, has failed to solve many of the problems associated with traditional riparian law. Because of the exemptions accorded large groups of users, regulated riparianism appears "piecemeal," incoherent, and protective of inefficient uses. Even in those cases where permits are required, agencies tend to favour existing or grandfathered uses, thus preventing reallocation from lower-to higher-value uses. In addition, because permits create rights that are temporary and often "use-specific," voluntary transfer markets have failed to develop. Agencies also often fail to institute conservation-oriented norms in their permitting decisions. Finally, agencies have yet to develop satisfactory methods for allocating rights during shortages. In short, permitting systems have not proved much more successful than their common law predecessor in addressing problems of conservation, efficiency, and equity.

It is thus time for eastern policymakers to reassess regulated riparianism. Indeed, some states are currently in the process of intensely reviewing their water allocation laws, while those that lack permitting schemes are likely to impose some mode of regulation in the near future. Most eastern water law scholars believe that the answer is more comprehensive management and greater centralization of control. Richard Ausness, for instance, calls for the abolition of exemptions for large-scale users, increased centralization and coordination of administration, and more detailed planning for shortages.

Joseph Dellapenna likewise characterizes less-than-comprehensive schemes as "an unseemly hodgepodge of requirements," and advocates the creation of "effective comprehensive planning mechanisms." But is a policy of greater centralization and increased managerial control necessarily the wisest approach? Are there alternative governance regimes that might achieve greater efficiency, higher levels of conservation, or a viable market in water rights? Drawing on the extensive common property literature, this Note argues that such alternatives are indeed available.

A wide body of scholarship in anthropology and institutional economics demonstrates that local users in a variety of contexts, from lobster fishing to irrigation, have been able to develop sustainable methods for managing scarce, common-pool resources, thereby overcoming a tragedy of the commons without the need for government regulation. This Note asserts that the organizational and institutional insights offered by common property scholarship are relevant to contemporary water management. The Note begins by suggesting that the common property model may not be such a remote concept in eastern water law after all. Indeed, close consideration of traditional riparian law reveals the outlines of the common property model embedded in the doctrine itself. Specifically, argues that the appurtenancy requirement within riparian law—restricting riparian rights to appurtenant owners and restricting usage to appurtenant land-bears distinct similarities to key features of common property regimes. Limiting rights to those owning riparian land functioned as a proxy for creating an exclusive community, while restricting usage to appurtenant land capped, in a rough way, total usage of the resource.

Drawing on the appurtenancy doctrine, as well as the common property literature, I urge greater attention to certain central principles of resource conservation and management—principles that the appurtenancy doctrine, properly understood, can be seen to support. I should be clear from the outset about the goals of such an interpretation of the appurtenancy requirement. Identifying similarities between the appurtenancy doctrine and common property regimes is meant to illuminate structural values that have proved durable in a variety of contexts. In particular, I emphasize the success of groups of users who are closest to a common-pool resource at managing against its scarcity. For critics of riparianism who may be skeptical of the wisdom of returning to the common law for inspiration, I would stress that it is appurtenancy's underlying principles—the principles of proximity to, and familiarity with, the resource at stake—and not merely the formal doctrinal rules that continue to bear relevance.

In fact, the rigorous requirements of the appurtenancy doctrine have been significantly modified in the last century and a half, and in many jurisdictions they no longer strictly govern. Part III examines case law reflecting this doctrinal shift and concludes that the decline of the appurtenancy requirement was the inevitable result of another key doctrine in riparian law: the reasonable use requirement, under which riparian rights are correlative and are limited only by the rights of other riparian proprietors. I argue that whereas the appurtenancy doctrine relied on a fundamental limiting approach to resources, the reasonable use doctrine, as interpreted and applied by the courts, relied instead on an inherently tolerant and expansionist ethic. Contrasting these two strands within riparianism serves several purposes: It provides a doctrinal explanation for the weakness of the appurtenancy requirement, it suggests a historical account of the law's failure to conserve, and it highlights a distinction—important for resource-management purposes—between rules that limit usage and those that balance users' interests.

It returns to the context of regulated riparianism, and argues that the appurtenancy and reasonable use doctrines have important implications for ongoing management of surface water in the East. I first note that existing regulatory schemes tend to replicate reasonable use doctrine and conclude that, in their current form, they are unlikely to effectively resolve scarcity problems. I then explore potential alternatives to comprehensive regulation that might better reflect appurtenancy's respect for those closest to a resource and their ability to conserve. While recognizing that small-scale community management may be unrealistic in the eastern water context, and a return to appurtenancy proper undesirable, I nevertheless urge consideration of institutional models combining greater local control with broader government oversight. Such "nested enterprises" can offer the advantages of both the commons and the state, while potentially mitigating their respective weaknesses; by incorporating appurtenancy into the modern regulatory context, they may thus provide the best opportunity yet to improve conservation, achieve greater efficiency, and ensure equity in eastern water usage.

## The Appurtenancy Requirement as Common Property Regime

At its core, riparian water law is about the entitlement of those bordering on a watercourse to certain use rights. The word "riparian" derives from the Latin ripa, which means the bank of a stream. Under the appurtenancy doctrine, a component of riparianism, only those who own property abutting the relevant body of water are entitled to access and use the water; in addition, water can only be used on adjacent parcels of land. This dual requirement has been subjected to substantial criticism, characterized as outdated and "anachronistic," accused of burdening urban and industrial expansion, blamed for "the development of uneconomical 'bowling-alley' parcels of land," and denounced for impeding the creation of marketable water rights and the increase in economic efficiency that many claim would follow. Indeed, overcoming the appurtenancy doctrine has been cited as "one of the principal motives [behind] the enactment of a regulated riparian statute." This Part reconsiders the appurtenancy doctrine in some detail and asserts that its underlying principles may prove more useful for resource management than its critics typically suggest.

The idea that the restrictions imposed by riparian law may have functioned as a common property regime has not gone unnoticed. Carol Rose, in arguing that common property regimes are more common than is generally recognized, briefly notes that "riparian law in the nineteenth century effectively turned river-bank landowners into participants in common property regimes... from which outsiders were excluded." Building on Rose's intuition, this Part fleshes out in more detail the various ways in which riparian law created, or at least attempted to create, a common property scheme, and compares this scheme to common property regimes in a variety of contexts. My goal here is to illuminate key principles for governing scarce resources.

### Access Restrictions in Common Property Regimes

Under traditional riparian law, only those who owned property bordering a watercourse were entitled to riparian rights. Conceptually, this ownership restriction limited access to water to a community of easily identifiable users. It is what Henry Smith would characterize as an exclusion rule, and it has proved central to conservation and effective management of common-pool resources in a variety of contexts. Exclusivity facilitates the development of social norms that favour conservation of scarce resources. James Acheson, in his study of Maine lobstermen, describes harbor gangs as exclusive groups in which membership is usually necessary in order to gain access to the resource. Indeed, some gangs require land ownership in order to gain entry, thereby paralleling riparianism's appurtenancy requirement. Because groups are limited in size, members are more likely to form "long-term, multistrand ties with one another," which in turn generate norms about "'proper' behaviour." In the fishing context, proper behaviour means eschewing overfishing: "There is a strong feeling that all members should have equal access to the resource.... Too much success, or success earned the wrong way... can generate envy, hostility, and gossip." Lobstering is recognized as a zero-sum game, and those who play too aggressively are disliked. Indeed, it is noteworthy that the most exclusive gangs, which place the greatest emphasis on longstanding kinship and community ties, are the ones that have "voluntarily agreed to limit the number of traps they fish."

Studies of Japanese mountain villages, Spanish huerta irrigation communities, Cree Indian fisheries, and Kwakiutl Indian potlatching practices likewise emphasize the relationship between exclusivity and sustainable resource usage. These communities developed detailed harvesting rules or fishing practices that prevented exhaustion of the resource at stake. As Bonnie McCay and James Acheson put it, territoriality—or "restricting access to a defined body of people who inhabit or lay claim to a particular territory"—provides "the basis for the development of more restrictive common property institutions.... If we can keep others out, it makes sense for us to do something about our own behaviour." In these communities, to which access is often strictly guarded, centuries-long traditions and "trial and error" experience breed conservationist norms.

Limiting group size also facilitates the monitoring and enforcement of usage rules. Small, enclosed communities benefit from "rich opportunities for mutual monitoring," which reduce the cost of finding violators. Both Acheson's lobster gangs and Spanish huerta irrigation systems display these traits. Moreover, because community members are repeat players, the threat of social sanctions and disrepute are often enough to produce adherence to usage rules. Fikret Berkes notes that among Cree Indians, "[e]ffective social mechanisms ensure adherence to rules which exist by virtue of mutual consent.... People who violate these rules suffer... social disgrace." Similarly, social prestige drives Kwakiutl Indians to engage in lavish potlatching practices, which D. Bruce Johnsen sees as a way to ensure territorial exclusion.

Finally, exclusivity allows community members to maintain a sense of control and autonomy. As one commentator has pointed out, "[P]roducers [in a common property situation] seem to oppose and flagrantly to violate any outside restriction on their practices, even when they would almost certainly regulate themselves if left in control." Barton Thompson's study of water institutions in the western United States illustrates the same point. In these institutions, where outsiders are regarded somewhat suspiciously, local users have been able to implement customized rules to conserve scarce water supplies much more successfully than has the state.

***The Ownership Restriction in the Common Law :*** Common property regimes thus succeed in conserving resources in part by excluding others. Exclusivity maintains user control, provides opportunities for extensive monitoring, and facilitates the development of restrictive use norms by those closest to the resource. As Anthony Scott and Georgina Coustalin have noted, restrictions on transferability of riparian water rights likewise seek to protect "the exclusivity of the right in the stream-wide group of users." I therefore turn to cases from the nineteenth and early twentieth centuries, which—while perhaps not always strictly adhering to the appurtenancy requirement demonstrate its inherently exclusionary tendencies. For instance, in cases dealing with grants of water rights to nonriparians, early courts held that such grantees were not entitled to full riparian rights. In Gould v. Eaton, for example, the defendants, who were nonriparians, had contracted for the right to place a tunnel on the grantor's riparian land in order to divert water. The California Supreme Court affirmed an injunction prohibiting the diversion and explained the extent of the defendants' entitlement under such a grant: "[It] gave [defendants] no right in the land which is adjacent to the stream, and they took by the contract no riparian rights in the waters of the stream...." While the defendants might have had a contractual right that was good as against the grantor, they had no rights as against other riparian proprietors. They were, in effect, excluded from the riparian community. Other courts have come to the same conclusion.

In one such case, Roberts v. Martin, the plaintiff was a gristmill operator, and the defendants owned nonriparian land near the plaintiff's parcel. In addition, the defendants purchased from an upstream riparian owner the rights to one of the springs on his land. The court held that their diversion of water violated the plaintiff's riparian rights. The court noted that the doctrine of reasonable use as between riparian proprietors did not include "'the case of an interference by a stranger, who, by any means, or for any cause, diminishes the flow of the water; for this also is wholly wrongful, and no question of the reasonableness of his action in causing the diminution can possibly arise.'" The court in essence announced a per se rule against water use by nonriparians, or "strangers," under which no such use could ever be found reasonable.

Perhaps more ambiguously, in Williams v. Wadsworth the Connecticut Supreme Court recognized the rights of a nonriparian owner based on his longstanding usage and acceptance within the riparian community. The plaintiff, a nonriparian, had purchased in 1851 from a riparian landowner "the right to build and maintain upon the land a dam," from which he diverted water via a pipe to his house. He used the water for domestic purposes, and also sold some of it to other town residents. The defendant, who owned a one-acre upstream parcel, began in 1881—thirty years after plaintiff had begun his usage—to divert water via "an acqueduct of his own... to his residence and premises," no part of which was riparian. The plaintiff's brief noted that after considerable expense, the defendant succeeded in supplying himself with water, which he used for domestic and culinary purposes, for watering his horses and cattle, for supplying a fountain or fountains, and for sprinkling his lawn and the street in front of his dwelling-house. A considerable quantity of the water thus diverted has run to waste, and none of it has been restored to the stream.

The court upheld the claim of the nonriparian plaintiff as against the riparian defendant, finding that he had acquired a prescriptive right through adverse use for the required statutory period. Given the uncertainties that often surround prescriptive rights in riparian law and the narrow scope courts generally give to such rights, it seems likely that the court considered other factors in finding the right. Indeed, the court took special notice of the fact that the plaintiff had secured from "all lower proprietors" a release by deed from any claims for his diversion.

Moreover, at the time that the dispute arose, the plaintiff had been using the water for approximately thirty years, while the defendant had been diverting for less than three years. The court thus seemed to find that the nonriparian plaintiff's usage—because longstanding and because explicitly accepted by the other riparian proprietors—was riparian, while the riparian defendant's diversion was not. Although the case might be read as an equitable exception to the ownership requirement, it can also be interpreted as confirming the community principle behind the appurtenancy requirement.

The release from liability that the plaintiff had sought and received from the other riparian proprietors, and his long-accepted diversion of the water, demonstrated that his usage had survived the community's scrutiny. In other words, his nonriparian status was outweighed by his proven entry into the web of interrelationships and monitoring that characterizes a common property regime. Finally, in a number of cases, courts granted plaintiffs injunctions against nonriparian diversions in part out of a concern that such nonriparian usage might ripen into a prescriptive right. The Roberts court, for example, noted that "[t]he unlawful act of defendants will, in time, ripen into an adverse right if permitted to continue. Equity affords complete and speedy redress in the premises." Such cases may have reflected the fear that a nonriparian user might forcibly gain access to the resource without properly entering the riparian community, again confirming the exclusivity principle behind appurtenancy.

The ownership component of the appurtenancy doctrine thus arguably gestures toward a key feature of common property management—limiting access to a group of users defined by their proximity to the resource at stake. A wide range of scholarship illuminates the ways in which such a restriction can assist in the conservation of a common-pool resource, and early case law reveals such dynamics guiding courts in their decisions. While exclusivity in itself is not necessarily admirable, other ideas central to appurtenancy—the ability of those close to the resource to develop sustainable practices and to monitor each other's usage, and the importance of user control over resource management—deserve attention in the establishment of any regime governing a scarce resource.

### The use Restriction

The other component of the appurtenancy doctrine is its use restriction, under which water may only be used upon riparian land. In this Section, I argue that appurtenancy and its sister doctrines operated more or less as a cap on total usage. I examine a number of cases demonstrating courts' sense that riparian land within the same watershed served as a rough proxy for the amount that could safely be withdrawn from the stream. I then consider usage restrictions in a range of common property settings to emphasize the importance of setting limits on withdrawal, as well as the competence of those most familiar with a resource in determining the appropriate level of withdrawal.

***The Use Restriction in the Common Law :*** The case of Williams v. Wadsworth illustrates the limitation principle at work. The plaintiff in Williams, though a nonriparian, had been diverting water, with the permission of all lower riparians, for approximately thirty years. When the defendant, a riparian, began diverting water to a nonriparian parcel, the plaintiff sued and was granted an injunction. The court's discussion of the extent of the defendant's riparian rights is telling:

Being a riparian owner the defendant has the right to consume water upon riparian premises.... But this use is to be confined to riparian land. This limitation... stands upon the necessity for a restraining rule in order to secure something for all, and upon the presumption that the brook will supply the absolute needs of as large an area as is usually held in riparian ownership.

The court then warned that "[i]f land not riparian may draw to itself, equally with land riparian... then land not riparian may take precedence of land riparian and deprive it of water for either man or beast." In short, the court's decision seems to rely on a rough estimate that the amount of water needed to supply riparian parcels generally approximates the amount of water that may safely be withdrawn without depleting the stream. Some readers will likely question how effective the appurtenancy requirement could be at confining total usage, given that a riparian landowner could in theory purchase larger and larger parcels of land and, under the unity-of-title rule that applies in most riparian states, could thereby exercise her riparian rights on an increasing area of land. In response, I would emphasize that the court's opinion in Williams is offered not to prove the conceptual limits of the use restriction, but rather to illustrate the underlying principles at work in the case law—conserving against scarcity and relying on the judgment of user-owners as to parcel size. In addition, and perhaps more importantly, the appurtenancy doctrine did not function alone, but rather operated in conjunction with other limitations on the total area of land that could benefit from riparian water rights.

First, many courts applied a "watershed limitation" in addition to the appurtenancy requirement; the conjunction of the two rules reflects a desire to limit withdrawal from the stream. In Anaheim Union Water Co. v. Fuller, the court held that "[l]and which is not within the watershed of the river is not riparian thereto, and is not entitled... to the use or benefit of the water from the river, although it may be part of an entire tract which does extend to the river." Although a parcel might be contiguous to a river, only that portion of the parcel within the river's watershed was considered riparian. The court went on to explain the logic behind this doctrine:

The principal reasons for the rule... are that, where the water is used on such land, it will... return to the stream, so far as it is not consumed, and that, as the rainfall on such land feeds the stream, the land is... entitled, so to speak, to the use of its waters.

Or as Richard Ausness explains, "The watershed limitation is based on the assumption that land beyond the watershed is outside the boundaries established by nature for riparian ownership...." In other words, the watershed requirement aimed to preserve the body of water in the stream by allowing it to be used only upon land that naturally returned such water to the stream. In this way, the appurtenancy and watershed rules worked together to reduce the effects of withdrawals. Some courts imposed even further limitations. For instance, the court in Town of Gordonsville v. Zinn limited riparian usage not only to land within the same watershed but also to land acquired in a single transaction. The plaintiff town owned a reservoir that supplied its residents with water. The defendant owned a very small parcel of riparian land—a twenty-five-foot strip—upstream from the reservoir, as well as a parcel downstream, upon which her house was located. Although the downstream parcel was also riparian, it had been acquired separately from the upstream parcel. The town sued the defendant when she began diverting water from her upstream strip to her downstream parcel, since water thus diverted bypassed the reservoir altogether. The court ultimately held for the plaintiff, enjoining the defendant from further diversion.

Putting aside for the moment the plaintiff's distribution of the water to nonriparian town residents, the court's opinion reveals a sense, similar to that displayed in the Williams opinion, that the size of riparian parcels will generally protect against depletion of the water source. Although the holding ultimately relied on the fact that the downstream receiving parcel was in a separate watershed from the upstream diverting parcel, and thus was nonriparian with respect to that diversion, the court nevertheless made sure to expound upon the single-ownership rule. Quoting from Farnham's treatise on water rights, the court took note of :

> *"the rule that all land must be regarded as riparian which is within the natural watershed of the stream, the title to which is in one owner and the boundaries of which have been established in accordance with the requirements of the conditions which will best serve the interests of individual landowners.... The watershed should certainly form a limit beyond which the riparian rights cannot be claimed; but there is a question whether or not that limit is not too wide.... Under this rule the boundary of riparian land is restricted to land... acquired by one transaction."*

Guarding against the possibility that land within a watershed, or separate riparian parcels held by a single owner, might "'recognize such a vast extent of territory as riparian land [so as to be] destructive,'" the court thus seemed to rely on the judgment of individual purchasers by using parcel size as a rough proxy for the area that could profitably be supplied by the stream in question.

### *Capping Mechanisms in Common Property Regimes*

Taken together, these cases suggest that the appurtenancy requirement and related doctrines functioned as a rough cap on withdrawal, by relying both on naturally arising drainage patterns and the judgment of riparian owners. The appurtenancy and single-transaction requirements in particular exhibited the law's faith that riparian proprietors would purchase parcels of sustainable size. I conclude this Section by examining a number of analogous property regimes in which usage rules function as capping mechanisms, restricting the amount that may be withdrawn from a common-pool resource. This comparison places appurtenancy's reliance on riparian owners within the context of a range of mechanisms adopted by appropriators to voluntarily restrict their own resource usage.

Multiple scholars have identified voluntary limits on total usage as a key factor in the sustainability of common property regimes. In describing the common grazing areas under the English open field system, for example, Smith notes that not only was access to common grazing areas restricted "to a small closed class of appropriators," but further restrictions on usage also applied. In particular, "stints," or limits on the number of animals that could graze on the commons, essentially functioned as a capping mechanism, since "one sheep cannot eat much more even if it is allowed to graze longer." Stints, like the appurtenancy requirement and its sister doctrines, served as "proxies for harmful activities"; the eating capacity of sheep, like the size of riparian parcels, inherently limited total usage. The success of the open field system, both in increasing productivity and in overcoming the potential tragedy of the commons posed by communal grazing areas, has been well-documented. Stints, which proved extremely "effective and widespread" in preventing overuse, were a central component of this success.

Other examples of voluntary restrictions on usage abound. Acheson credits lobster gangs' self-imposed reductions in fishing efforts—in particular, voluntary limits on the number of traps, which necessarily limited total catch —for "both economic and biological benefits," including greater lobster density and higher lobstermen profits. Swiss villages have sustained a communal alpine economy since the thirteenth century by requiring that officials first "mark the trees ready to be harvested," which allows for "a careful assessment of the condition of the forest." In the extensive common lands regulated by Japanese villages for centuries, "detailed authority rules specif[ied] in various ways how much of each valued product a household could harvest from the commons and under what conditions." And, perhaps most pertinent to the appurtenancy context, more than one huerta irrigation system has limited water rights to certain areas of land; designations of some parcels as irrigated and some as dry have lasted for centuries.

Examining the use restriction of appurtenancy doctrine in light of the practices of common property regimes allows us to make two observations. First, from a resource standpoint, the restriction, while no doubt often inefficient and inequitable in confining water usage to adjacent land, may nevertheless have stemmed from a quite rational desire to limit withdrawal by users of a scarce resource. Appurtenancy and its sister doctrines may have tried to cap usage by drawing on the proxies of adjacency, watershed boundaries, and parcel size, just as medieval peasants relied on sheeps' digestive systems, and Maine lobstermen relied on the size of standard fishing traps. Second, as a structural matter, both appurtenancy and successful common property regimes seem to have relied on the intimate knowledge of proximate user-owners in determining the appropriate level of withdrawal. For common-pool resource management, both insights—about the need to cap withdrawal and the expertise of those closest to (and thus most familiar with) a resource in setting such caps—are essential.

## Reasonable use and the Decline of the Appurtenancy Requirement

This Note began with the premise that, in light of continuing water shortages, eastern states need to reassess current water regulation and develop more sustainable resource-management practices. Resisting the preference exhibited by most commentators for comprehensive and centralized regulation, I instead turned to an unlikely place—the appurtenancy doctrine—and suggested that appurtenancy contained certain structural features that have proved successful in governing and conserving scarce resources in a variety of common property contexts. A complete account, however, must recognize that the success of riparianism in restricting access and usage has been, at best, quite limited.

Though modern treatises still define riparian rights as those "concerning the relation of the owner of the bank of a stream to various features of the stream," most commentators recognize that the appurtenancy doctrine—the feature that gives riparianism its very name —is today subject to numerous exceptions. "[E]very state," notes James Christman, "allows some means of transferring riparian rights apart from the land." Granted, the law regarding these transfers varies from jurisdiction to jurisdiction and remains far from clear: Some states recognize conveyances, though whether such conveyances are valid only against the grantor, or against all other riparians, is a matter of some dispute; a number of states recognize the right to lease riparian water rights; and a fraction of all states recognize prescriptive nonriparian rights.

Nevertheless, the general trend has been away from appurtenancy and toward increasing nonriparian use. Recent reformulations of the law likewise recognize the need for nonriparian rights. The Restatement (Second) of Torts does not consider "classification of [a use of water] as riparian or nonriparian" to be controlling, and the influence of this view is seen in court decisions—most notably Pyle v. Gilbert —that recognize nonriparian rights. And finally, the permitting systems and regulated riparian regimes that increasingly govern surface water in eastern states also recognize nonriparian rights as permissible. Why did the appurtenancy doctrine never really take hold, and what relevance does this have for modern resource management? In this Part, I offer two accounts. I examine cases in which courts made exceptions to the appurtenancy requirement; this doctrinal analysis leads me to conclude that the reasonable use doctrine that increasingly governed riparian water law during the nineteenth century inherently allowed for expansion of the riparian community and the acceptable set of riparian uses. Section B offers some broader historical reasons for the failure of the appurtenancy doctrine, emphasizing the cavalier approach to natural resources that characterized the United States during the era of industrialization. In the next Part, I draw on these conclusions and suggest that the tolerance for increased usage inherent in the reasonable use doctrine continues to have significance today.

### Reasonable use as Doctrine

Under the reasonable use doctrine, "each owner of riparian land is permitted to use the water in a waterbody, regardless of the effect the use has on the natural flow, so long as each user does not transgress the equal right of other riparians to use the water." The rise of reasonable use doctrine during the nineteenth century, replacing the more stringent natural flow theory of riparian lights, has been the subject of considerable scrutiny. The argument here is that reasonable use, at least as framed by nineteenth-century courts, was a doctrine inherently tolerant of expanding uses of water; as such, it operated in tension with the appurtenancy doctrine, which I have characterized as inherently restrictive.

The classic statement of riparianism in American law is Justice Story's opinion on circuit in Tyler v. Wilkinson. In that case, the lower riparian plaintiffs claimed that upstream proprietors were entitled only to water that the plaintiffs did not need. The court, however, rejected this suggestion, holding that "[t]here may be, and there must be allowed of that, which is common to all, a reasonable use." While the opinion seemed to hold fast to natural flow theory in some respects, it also acknowledged what Morton Horwitz has called "the utilitarian criterion of valuable use [as] the ultimate source of legal rules." Specifically, Story rejected the idea that riparianism would allow for "no diminution whatsoever... for that would be to deny any valuable use." The Tyler case, while still nominally adhering to natural flow doctrine, thus opened the door to a balancing approach to riparian rights that would take into account "valuable use."

Further elaboration of the reasonable use doctrine has only added to its malleability. In the 1883 case of Red River Roller Mills v. Wright, for example, the Minnesota Supreme Court offered this definition of the factors to be taken into consideration in a reasonable use analysis:

In determining what is a reasonable use, regard must be had to the subject-matter of the use; the occasion and manner of its application; the object, extent, necessity, and duration of the use; the nature and size of the stream; the kind of business to which it is subservient; the importance and necessity of the use claimed by one party, and the extent of the injury to the other party; the state of improvement of the country in regard to mills and machinery, and the use of water as a propelling power; the general and established usages of the country in similar cases; and all the other and ever-varying circumstances of each particular case bearing upon the question of the fitness and propriety of the use of the water under consideration.

The reasonable use analysis thus became open-ended and case-specific. Such a utilitarian approach, concerned as it is with comparing social costs and benefits, is inherently open to expansion; it allows—indeed, it encourages—courts to take into account the economic value of the infringing user's activity. As Smith notes, such tort-based rules do not protect rights, but rather focus on "the efficient use of resources. Scarcity over resources is reduced to bilateral conflicts over use." The key considerations are "maximizing value" and "optimal resource allocation," rather than protection against global scarcity. A reasonable use regime is governed by considerations of "social values and needs" and "general rules of permissible and impermissible human conduct"; as such, it is inherently "fluid and socially responsive."

Finally, consider the focus on actual injury in the reasonable use inquiry. As Justice Story put it in Tyler, "The true test of the principle and extent of the use is, whether it is to the injury of the other proprietors or not." Courts often cite the doctrines of de minimis non curat lex and damnum absque injuria in rejecting plaintiffs' claims, finding damage nonexistent, or simply too small to justify a remedy. This rule, too, fits with an expansionist interpretation of reasonable use: Defendants will not be enjoined, even if their activities violate established rules of riparian use, unless plaintiffs can actually show injury.

Reasonable use's tolerance for expanded usage is perhaps most apparent in those cases allowing exceptions to the appurtenancy requirement. In these decisions, reasonable use essentially swallows up the appurtenancy doctrine. In Elliot v. Fitchburg Railroad, for instance, the defendants purchased a riparian parcel upstream from the plaintiff's land and diverted water to their depot to furnish locomotive engines. Although the diversion violated the appurtenancy rule, the court held the defendants "not liable" because the plaintiff had not "suffered actual perceptible damage." Reasonableness, it held, was "a question of degree," not the absolute rule appurtenancy would impose.

Similarly, in Jones v. Conn the court allowed the defendant to irrigate nonriparian land and power a gristmill over the complaint of the plaintiffs, who used the water to irrigate riparian parcels. In opposition to the theory of appurtenancy, the court noted that "whether a proposed use is... reasonable, does not depend so much upon the area of the land... or the place of use, as upon the effect it has upon... the other proprietors." Here, the "amount of water taken... was not sufficient to materially injure the plaintiffs," so the defendant could not be deemed a "wrongdoer."

Even more extreme is the Oklahoma Supreme Court's statement in Smith v. Stanolind Oil & Gas Co. The plaintiffs sought to enjoin Stanolind from obstructing water flow and taking the water for nonriparian purposes. The court emphasized that nonriparian use "standing alone, [is] no criterion of the reasonableness of the use.... [A] riparian owner has the right to make any use of water... possible, so long as he does not inflict substantial or material injury [on other riparians]." Doctrinally, one can therefore link the decline of the appurtenancy doctrine—or perhaps more accurately, the refusal of courts to insist upon its strict application—to the rise of reasonable use. The appurtenancy doctrine, which we might characterize as a resource-focused limiting rule, conflicted with the user-focused balancing rule of reasonable use. This account is thus somewhat in tension with Rose's description of the reasonable use doctrine:

> [R]iparian law in the nineteenth century effectively turned riverbank landowners into participants in common property regimes for particular rivers, from which outsiders were excluded (no interbasin water transfers) and insiders were expected to use the common resource "reasonably"—in effect, in a way that was compatible with equal correlative use by all the other riparian owners.

In other words, Rose suggests that both place-of-use restrictions and the reasonable use doctrine were a part of the common property regime. The view offered here, however, is that reasonable use actually worked in opposition to and defeated the restrictive common property elements of appurtenancy. While reasonable use initially may have governed a community of insiders, the doctrine's tendencies to treat uses and users "reasonably," to consider the social value of activities, and to decide in favour of efficiency and productivity, would inevitably lead to expansions of the community as well as the range of accepted uses.

While the Tyler test of injury initially looked only to the impact on, and competing claims of, other riparian proprietors, its principle of balancing and value maximization would eventually require consideration of the interests of outsiders. In short, reasonable use sooner or later tended to undermine appurtenancy. This account of reasonable use versus appurtenancy allows us to draw a conceptual distinction among different types of what Smith calls governance rules—a distinction not fully apparent from Smith's analysis, but important for resource-management purposes. Under Smith's taxonomy, governance rules involve the "specification of proper activities" for which a resource may be used.

Given this broad definition, both appurtenancy's use restriction and the reasonable use doctrine would qualify as governance rules, since both speak to appropriate usage: Appurtenancy specifies the parcels of land on which water may be used, while reasonable use (at least in theory) specifies the purposes for which it may be used. Smith's focus is on the relative information costs of different rules, and he thus differentiates between governance rules of greater and lesser specification. He does not, however, distinguish between rules that limit total usage and those that merely aim to balance conflicting users' interests. He sees governance, like exclusion, as an "alternative method of preventing dissipation," without recognizing that certain governance rules, such as the reasonable use doctrine, may actually facilitate increased usage of a resource. Smith notes that common property regimes often involve both exclusion rules and "rules of proper use among [group members]," which may "promote conservation and sustainability of common-pool resources."

But he does not acknowledge that some "proper use" rules emphasize limitation, while others may focus on value maximization—a distinction that is at the heart of riparian law's shift from appurtenancy to reasonable use. Smith does briefly address the transition from natural flow theory to the reasonable use doctrine, but asserts only that "exclusion tends to give way to governance as pressure on the resource has increased," without analysing the implications of reasonable use for a resource under pressure. His analysis thus misses the key difference among governance rules that this Note has emphasized in the two riparian regimes. Examining the limited success of the appurtenancy doctrine in relation to the reasonable use doctrine, and analysing the cases in which courts allowed the former to give way to the latter, one can see an important conceptual distinction—limiting versus balancing—among various regimes aiming to conserve scarce resources.

## Reasonable use as Historical Development

The rise of reasonable use, however, is not an entirely satisfactory explanation of why appurtenancy failed to take hold. Indeed, doctrinal analysis simply begs the question: It explains what judges did, but not why they did it. In this Section, I briefly offer some suggestions as to why judges turned away from appurtenancy and constructed an ever-broadening notion of reasonable use.

Morton Horwitz's legal realist account holds that the rise of reasonable use and fall of natural flow was part of what he calls "the transformation in the conception of property" that occurred during the nineteenth century. The reasonable use standard was "an open-ended formula through which common law judges could implement their own conceptions of desirable social policy." What I, somewhat agnostically, as the value-maximizing tendency of reasonable use is viewed by Horwitz in more political terms, as "central... to the needs of industrial development... in the nineteenth century" and "essential to economic improvement." Rose summarizes Horwitz's explanation succinctly: "Legal developments [in water law] gave a kind of common law subsidy to capitalist developers, allowing them to inflict some injury upon weaker and less enterprising folk without having to pay for the consequences." Theodore Steinberg, an environmental historian, agrees: "The drive to promote industrial growth had, in the process, reduced water to an instrument in the service of economic change. Water by the early years of the nineteenth century was well on its way to becoming a simple utility, an abstraction employed to suit economic ambitions.... a commodity."

Other, less political explanations are available, and they complement Horwitz's account. As riparianism moved beyond New England, into southern and western parts of the country, it encountered regions where water was used less for power—an instream use that would not have found the appurtenancy doctrine unduly restrictive —and more for consumptive uses such as irrigation and farming. In these areas, a more tolerant attitude toward withdrawals by nonriparian owners for use on nonriparian land undoubtedly seemed more attractive. Moreover, once steam became a major source of energy, and the mill economy of the Northeast declined, "courts began to appreciate the fact that it would be efficient to increase the consumption of many streams and they began to develop a law of consumption through the reasonable use theory." In short, both economic and geographic expansion of the nation likely gave impetus to doctrinal change.

Finally, the need to expand the acceptable uses of water went unchecked by any sense of scarcity. Quite simply, the humid eastern states did not face the same shortages as western states. Courts lacked a pressing reason to adhere to the logic of conservation and restriction embedded within the appurtenancy doctrine. With water plentiful and the economy expanding, the dominant ethic of industrialization demanded the fullest use of natural resources. In such a climate, courts found little need to insist on ownership or use restrictions, and plenty of reasons to tolerate expanded access and withdrawal.

This Part has offered two explanations for the weakness of the appurtenancy requirement in riparian law. As a doctrinal matter, appurtenancy and its restrictive attitude conflicted with, and succumbed to, the more tolerant balancing approach of reasonable use. As a historical matter, appurtenancy's values of limitation and conservation did not fare well under the pro-development, efficiency-maximizing policies of the nineteenth and early twentieth centuries, nor did they seem necessary in light of the seeming abundance of water. This comparison of the reasonable use and appurtenancy doctrines proves helpful as we return, in the next Part, to regulated riparianism.

## The Continuing Relevance of Appurtenancy's Principles

Having considered in some depth the two basic doctrines of riparianism—appurtenancy and reasonable use—as well as the features of common property regimes that have enabled users to sustain and conserve scarce resources, we are now able to draw several conclusions to aid us in our evaluation of regulated riparianism: First, restricting access to a resource to its closest users can facilitate the development and enforcement of conservation norms. Second, because users are generally resistant to external oversight, placing management authority in their hands will often increase the likelihood that they will adopt effective resource usage practices. Third, because users will generally have greater familiarity with the resource and its changing dynamics than government actors, they are often better able to determine appropriate levels of usage and to limit their withdrawals accordingly. Finally, regimes that rely on a value-maximizing, balancing approach are less likely to protect scarce resources against growing demand than regimes that take a more restrictive, limiting approach. With these observations in mind, I offer a critique of current water law regimes in Section A. While a full-scale policy proposal is beyond the scope of this Note, I do suggest, in Section B, a number of ways in which the lessons of appurtenancy and common property management may be applied to water law today.

### *Reasonable use Redux*

Begin by drawing attention to the survival of the reasonable use doctrine under current regulatory schemes. Most statutes list a set of factors for agencies to use in deciding whether or not to grant a permit; such lists closely resemble the open-ended inquiries conducted under the reasonable use doctrine in cases such as Red River Roller Mills. The Regulated Riparian Model Water Code, for example, provides a list of nine factors to be taken into account by the permitting agency, including "the economic and social importance of the proposed water use and other existing or planned water uses sharing the water source." Such lists "can only reinforce the sense that the substantive basis for decision remains much the same as in traditional common-law litigation under the reasonable use theory of riparian rights." While advocates of riparian statutes stress that such decisions are made ex ante, rather than ex post, I would question the extent to which moving the reasonableness inquiry from an ex post judicial decision to an ex ante administrative determination can counteract the tendency of such balancing tests to permit increased usage, instead of protecting scarce resources. As one scholar comments, "[I]n permit-based jurisdictions, allocative standards make little effort to distance themselves from their riparian ancestors."

Supporters of regulated riparianism may note that every permitting statute requires the administering agency to develop long-term plans to protect the source for future needs, and that almost all permitting statutes also "require the agency to define or protect minimum flows." But the existence of such statutory mandates does not ensure protection of water resources. Long-term plans, for instance, are often developed according to a list that, again, echoes the reasonable use inquiry; they are thus unlikely to result in greater conservation. Furthermore, statutory mandates to protect minimum flows have proved, in practice, difficult to administer, given "traditional water management" values and the "considerable discretion" granted administering agencies. In Arkansas, for example, where a regulated riparian statute provides for the protection of minimum stream flows, "[t]he process of setting minimum flows has proven highly political, [resulting in] considerable modification (and reduction) of the proposed original levels of protection." A series of recent agency and court decisions in Vermont has severely limited the ability of the state's permitting agencies to use the public trust doctrine to protect minimum stream flows. Lack of strong enforcement of minimum-flow protections, combined with the persistence of reasonable use inquiries in permitting decisions and long-term planning, may thus indicate that regulated riparianism will fare no better at conserving water supply than its common law predecessor.

### Local-global Possibilities

Rather than press for greater centralization and control while adhering to reasonable use inquiries and multifactor tests, regulators should take a step back and consider potential alternatives to the comprehensive management model. Indeed, common property scholars have extensively documented the disastrous consequences—in both ecological and human terms—that can result from wrongheaded state intervention. In light of the analysis offered in this Note, I would urge serious consideration of regulatory responses that attend to scarcity in a way that incorporates the core principles of appurtenancy. In other words, while regulated riparianism has until now included primarily the reasonable use strand from common law riparianism, I would instead emphasize the appurtenancy strand.

Some readers will no doubt question the feasibility of returning to the strictures of the appurtenancy doctrine. Indeed, confining water usage to adjacent bodies of land, for the benefit of the owners of those parcels, seems neither a realistic nor a desirable goal for contemporary water management. I would suggest, however, that appurtenancy need not be conceived of solely as a physical restriction. Instead, I propose here an understanding of appurtenancy as a structural, organizational concept for resource management. It is this reconceptualized notion of appurtenancy that may prove valuable for policymakers today.

### *Reconceptualizing Appurtenancy*

The traditional understanding of appurtenancy focused on physical proximity: Only owners whose property abutted the water had rights to it, and only their adjacent land could benefit. But as the cases reveal, and as comparisons to common property scholarship highlight, the doctrine was not solely about physical proximity, but also contained certain structural principles of resource management. Limiting riparian rights to adjacent owners excluded "strangers" and thus defined a group of users who could enforce conservationist norms of behaviour. Restricting usage to appurtenant parcels demonstrated a reliance on those most familiar with the resource to determine appropriate levels of usage. In other words, appurtenancy doctrine expressed a commitment to vesting resource management in those closest to, and sharing an interest in, the waterbody. In the eighteenth and nineteenth centuries, this structural commitment coincided with physical proximity; adjacency functioned as a proxy for defining the most knowledgeable community of users. But in a twenty-first-century world in which water consumption needs are vastly different, we should perhaps de-emphasize the physical requirements associated with appurtenancy and focus instead on its delegation of governance rights to those most intimate with the resource, whether physically proximate or not.

Under this definition, appurtenant users would no longer necessarily be those with property bordering a given resource, but would rather be defined as those within a given region using a resource in a similar way. Membership in a community of users would be conceived of not in terms of adjacent land ownership, but rather in terms of the particular use—whether irrigation, recreation, or municipal water supply. Appurtenancy, on this understanding, would vest management authority not in those physically proximate, but in those functionally proximate—both to the resource and to a like group of users.

Putting appurtenancy to work in a contemporary water regime would thus mean recognizing the competence of communities of users in effectively governing scarce resources. At first glance, the workability of such a governance scheme in the eastern water context might seem dubious. After all, the examples of community management generally involve close-knit groups that are able to exclude outsiders because of their small size, and whose ability to conserve resources depend largely on longstanding social ties. While such community-based regimes might work in the context of small streams or rivers, they seem implausible when it comes to larger watercourses that may span not only great distances, but also multiple jurisdictions. Moreover, unlike the irrigators in huerta communities or the farmers in medieval England, the heterogeneity of water users in eastern states makes traditional collective action seem unlikely to succeed. And while common property management has proved successful at addressing conservation concerns, its exclusionary nature and resistance to outside interests might not sit well with those worried about efficiency and equity.

In response to such concerns, I would first suggest that appurtenant groups need not be limited to homogeneous communities of individual users; users drawing on the same body of water for similar purposes could be considered appurtenant, even if they were institutional entities. Second, I would emphasize that small-scale community-based management and comprehensive government regulation are not the only options. We may do better if we avoid "the familiar trichotomy of private, communal, and state ownership," and instead pay more attention to the possibilities of localstate co-management. As Oran Young suggests, rather than "focusing exclusively on finding the right level or scale at which to address specific problems.... the key to success [may instead lie] in allocating specific tasks to the appropriate level of social organization and then taking steps to ensure that cross-scale interactions produce complementary rather than conflicting actions." Indeed, much recent work in common property scholarship and institutional economics more broadly has been directed toward the possibilities of local-state co-management. Thus, just as the Note drew on the first wave of common property scholarship to illuminate our understanding of the appurtenancy requirement, I now turn to more recent work in the area to inform our regulatory policies going forward. Specifically, I consider what Elinor Ostrom has called "nested enterprises," in which "[a]ppropriation, provision, monitoring, enforcement, conflict resolution, and governance activities are organized in multiple layers." Under such a model, respect for appurtenant communities could be embedded within a larger-scale government system.

While nested enterprises may sound intriguing in theory, readers may wonder whether such efforts work in practice, and how they would work in the context of eastern water law. In the remainder of this Note, I therefore describe successful examples of cross-scale management. I also suggest reasons to believe that such regimes may provide the advantages of both the commons and the state, while mitigating, rather than exacerbating, their respective weaknesses. I conclude by arguing that multilevel governance may be particularly well-suited to the dilemmas facing eastern water regulators. In particular, nested regimes that embed appurtenant groups within an administrative scheme may more effectively address the three major concerns conservation, efficiency, and equity—that surround allocation of our water supply.

### Nested Enterprises: Some Models

Numerous instances of successful nested enterprises exist. This Subsection explores four models and considers in each instance how norms of usage and conservation developed by appurtenant groups can be effectuated within a larger government system. These examples highlight the ways in which nested enterprises may yield the advantages of both the commons and the state, while mitigating the weaknesses of each. One nested enterprise model, which we might call the reverse command-and-control model, utilizes state authority, expertise, and administrative resources to enforce usage roles developed by appurtenant groups to protect against scarcity. For example, the Maine lobster gangs have, in recent years, formed links with state government, which in turn aids them in the enforcement of their voluntary restrictions on fishing. Acheson notes that the inhabitants of Monhegan Island have "persuaded the legislature to pass a law forbidding fishing in Monhegan waters from June 25 to January 1," thereby making official a customary closed season that the islanders have observed since the early 1900s. Fishers from Swan Island likewise convinced Maine's commissioner of marine resources to enforce a self-imposed limit on the number of traps surrounding the island, A similar example comes from the Aleut community in Alaska, where the community controls access to the territories, and thus can control the level of fishing, but has also "made the state limit fishing effort in adjacent territories where the local stocks might be intercepted by outsiders."

Yet another example can be found overseas, in the state of Kumaon in India, where centralized rules create "a framework for the management of forests rather than a defining straitjacket," and at the same time establish "a domain of relatively autonomous action and rule making in which local residents and their representatives can operate." Specifically, local authorities "make rules and enforce them," "facilitate some kinds of actions... and restrict others," meet frequently to modify rules, and "create monitoring and sanctioning mechanisms." The state government supplements these efforts by supervising recordkeeping, coordinating harvesting, providing technical assistance, and aiding in rule enforcement. As these examples demonstrate, reverse command and control takes advantage of the greater skill of local users in developing rules of conservation, and places those rules under the enforcement authority of government.

Another nested governance model that has gained attention recently combines market-based tradable allowances with community management. Under tradable allowance schemes, governmental regulators in effect place an upper limit or cap on the total quantity of a given resource that is to be available for use.... The regulators then divide the capped total into individual allowances. Henceforth they require all resource users to purchase or trade for whatever allowances they use.

In a nested enterprise model, quotas are allocated to appurtenant communities, rather than individuals or individual entities. In Alaska, for example, some of the walleye pollock quota was allocated to native communities; similarly, New Zealand transferred forty percent of its fishery quota to the Maori. Such schemes incorporate groups of users-proximate both to each other and to the resource—into a larger regulatory market, thereby protecting community interests while maintaining the advantages of a centralized decisionmaker for setting and managing overall goals. In these regimes, as compared to the reverse command-and-control model, the state does not undertake enforcement of specific usage restrictions developed by the user community. Instead, it grants an appurtenant group of users a block of rights that is presumably large enough to allow the group to maintain a sustainable level of withdrawal, leaving the group to implement and enforce its own conservation norms. At the same time, the tradable allowance scheme permits user groups to operate within a broader market of uses and users. The existence of a wider market in standardized rights can facilitate transfers from lower-to higher-value uses, thereby increasing the efficiency of resource usage.

The creation of standardized rights can also facilitate coordination across different appurtenant groups—functioning as a kind of common language and in some instances leading to larger-scale conservation efforts. A third model—what we might call the horizontal governance model—comes from voluntary associations of groundwater users in Southern California. These nested enterprises should be particularly persuasive not only because they involve water as opposed to fisheries or forest management, but also because they demonstrate the viability of such a model among institutional, as opposed to individual, users. The parties involved here are not members of an Indian village, or an Aleut fishing community, but include municipalities, utility companies, and private mutual corporations.

Elinor Ostrom documents the rise of "a set of institutions [established by water producers] to manage a series of groundwater basins located beneath the Los Angeles metropolitan area." Faced with the prospect of saltwater incursions, dropping aquifers, and uncertain rights, appropriators began to engage in a "pumping race," a classic example of a tragedy of the commons. During the 1940s and 1950s, many believed that the race would result in the destruction of the basins, with dire consequences for the water supply in southern California. But instead of succumbing to the tragedy, water producers entered into a series of negotiations, through which they were able to form institutional structures that combined public and private governance, to set restrictions on withdrawals, and to institute conservation measures. Nested within the public arena, private actors were able "to impose constraints on themselves."

A few features of these settlements are particularly noteworthy. First, the disputes arose among functionally appurtenant users—claimants with shared interests in the same body of water. Second, in coming up with a solution, it was important to the users that they maintain some measure of management authority; they did not want to turn decisions over to large-scale agencies, "for fear that they would lose control of the decisions being made and might end up worse off." In short, they exhibited the same desire for autonomy and self-governance as members of the common property regimes. Third, while users maintained decisionmaking control, they did so within a larger governmental regime; the institutional solution was neither deregulation nor centralization, but rather "polycentric" systems of shared private-public governance. While the first two models of nested governance are largely vertical and thus somewhat hierarchical, with the state at the top and user groups at the bottom, this model vests management authority in a more horizontal, or multipronged, manner. Fourth, the costs of monitoring and enforcement were shared by the users (two-thirds) and the state (one-third) And finally, the negotiations among parties led to a better clarification of rights, which in turn allowed a market to develop, leading to a transfer of rights to those using them at "a higher value."

The Ostrom study allows us to draw two important conclusions. First, appurtenant groups can be defined to include institutional users—here, the appurtenant groups were voluntary associations made up of both individual and institutional user-appropriators—who maY prove just as capable of developing and enforcing conservation norms as individuals within a small community. Specifically, institutional appropriators restricted their water usage and monitored each other's withdrawals, just as farmers in huerta villages restricted water usage to certain parcels of land and kept an eye on their neighbours' withdrawals. Second, a nested enterprise model in which management is "polycentric" rather than centralized, and in which appurtenant groups of organized users actively participate can produce an allocation of tradable rights that is efficient and stable, as well as conducive to conservation. As clarification of rights has proved difficult under regulated riparianism, and a market in permits has failed to materialize, the achievements of the entities Ostrom describes are particularly noteworthy. In short, the horizontal governance model seems to have had more success in resolving efficiency and conservation problems than have eastern regulatory systems.

A fourth model—what we might call the umbrella model—is illustrated by local water institutions in the western United States. Mutuals and water districts are the retail distributors of most domestic water and much agricultural water in western states. Mutuals are private nonprofit corporations whose customers are also shareholders; water districts, on the other hand, are governed by elected boards, like other local government units. "Both types of institutions engage in a broad set of activities, including obtaining and storing necessary water supplies, transporting the water to their service areas, and distributing it to their members." For our purposes, mutuals and districts may be analogized to appurtenant groups (though the water they consume may come from a distant reservoir or dam), as they allow groups of local users to access water jointly, thereby taking advantage of economies of scale, and to maintain control over allocation within the group.

These institutions have developed into multilayered regimes, thus exemplifying a more complex, cross-scale kind of nested enterprise than the models already discussed. Mutuals and water districts may obtain their water from a larger umbrella agency acting as a wholesaler, or they may sit within larger umbrella institutions themselves. All of these institutions ultimately sit under the general authority of the legislature. Several features of these institutions are familiar. For instance, whereas users have generally been resistant to state proposals for conservation, they have been receptive to conservation programs implemented by local institutions. These programs have been able to reconfigure groundwater rights; impose well-spacing rules; institute pump taxes and import projects to optimize use of ground and surface water; and establish loan programs, conservation rebates, and, in some cases, tiered pricing structures. In short, appurtenant groups of users have, here as elsewhere, voluntarily recognized the need to conserve, and they have advanced creative methods to meet that need, but in a way that emphasizes local control.

Thompson also describes at length the failure of state governments to facilitate voluntary transfers: "Legal restrictions on formal water transfers" pose insurmountable transaction costs for many small water users...." In response, users "have employed local institutions to obtain or exchange water." Institutions provide members with "largely interchangeable water entitlements," allowing the development of standardized markets; members of institutions are also likely to be familiar with each other's needs; and many institutions have constructed extensive physical infrastructure that facilitates the transfer of water from one user to another. Finally, institutions have proved adept at allocating water during times of scarcity: They can "balance their service areas, water supplies, and storage capabilities.... [They] can also more readily calculate and enforce pro rata allocations." In the language of appurtenancy, physical, social, and functional proximity among users facilitates efficient trade and reallocation within the group. Of course, mutuals and, to an even greater extent, districts can exhibit some of the less savory features of appurtenant communities as well. While they vary in size, and while some encompass a large geographic area and a range of users, their success seems due in part to the homogeneous—and sometimes insular—interests of their membership. Many mutuals are "enlivened with a dollop of community spirit." While such community spirit may make informal transfers between members less cumbersome than statutory transfers and encourage the adoption of stronger conservation measures than those imposed by the state, internal community norms also reflect the exclusionary tendencies and insider-outsider mentality that an appurtenant group of users can exhibit. In particular, Thompson notes the resistance of some local institutions to interjurisdictional transfers, such as "ag-urban" trades. Such resistance can often be the result of "parochical self-interest that is inconsistent with broader societal good." Importantly, however, Thompson does not advocate complete centralization of control as the solution to this problem, and he is wary of stripping too much authority from local institutions and losing the benefits of appurtenancy:

> Despite the potential advantages of creating larger water institutions... attempts to enlarge or consolidate current institutions will inevitably encounter numerous problems and limitations....... Legislatures therefore may do best by merely encouraging institutions to examine the advantages of forming larger units, and then making it as easy as possible for them to do so.

This approach might best be described as more nesting—increased usage of intermediate, umbrella institutions that can facilitate interregional trades among various sets of appurtenant groups, and increased legislative support for such institutions. While Thompson recognizes that more heavy-handed state oversight may in some instances be necessary, his preferred solution is neither purely regulatory nor purely market-based, but is rather institutional. In the multilayered regime he proposes, larger associations can overcome obstacles to market trade that currently exist by producing internal economies of scale and internalizing the costs imposed by one group on others.

***Some Policy Suggestions :*** Having reconceptualized appurtenancy as a principle of functional rather than physical proximity, and having determined that appurtenancy can function within a larger-scale nested regime, we can now begin to ask how such a regime might work in practice in the context of eastern water. The four models outlined above—reverse command-and-control, market-based tradable allowances allocated to communities, horizontal governance, and increased nesting through umbrella organizations—may provide policymakers with a starting point. Berkes, using different terminology, similarly suggests a range of ways in which states can facilitate cross-scale management, including "legitimization of local institutions," enabling legislation that gives authority to "locally devised rules," capacity building "at all levels," and "institution building." What "legitimization" or "institution building" will mean in practice will depend on the particular institutional structure in place in each jurisdiction, and the types of user associations that exist or that might be formed. While a comprehensive policy proposal is beyond the scope of this Note, I would like to offer a few preliminary thoughts as to how policymakers might incorporate appurtenancy into eastern water management.

Given the efficiency and equity concerns posed by purely local organizations, state regulatory agencies should retain ultimate authority over the allocation of water rights. The question, then, is how states might recognize appurtenant groups and include them—formally or informally—in water management decisions. Perhaps the simplest option would be to encourage the development of voluntary user associations, without providing such groups any formal legal recognition. Such an arrangement would allow appurtenant groups to form through self-identification, perhaps utilizing industry or interest groups already in existence. States could offer encouragement in several ways. As Ostrom's study of groundwater basins in southern California demonstrates, provision of a public forum for users can facilitate negotiation, information-sharing, clarification of rights, and adoption of voluntary constraints by users. The state might further support voluntary user action by offering assistance with information-gathering (recall the recordkeeping function provided by the state of Kumaon to local authorities) and the enforcement of any voluntary agreements among users (for example, enforcement of trap limits in Maine lobstering waters, and the provision of one-third of the costs of a monitor for groundwater basin withdrawals in southern California).

This approach could be characterized as capacity-building or institution-building, within Berkes's taxonomy, and would in some respects parallel the reverse command-and-control model delineated above. The primary objective would be to aid in the development and enforcement of conservation norms developed voluntarily by those most familiar with the resource.

Beyond such informal facilitation, state regulators could explore a variety of ways to offer appurtenant groups more formal legal recognition. One possibility that could work within a modified version of existing permitting systems would be to issue blocks of permits to groups of users rather than to individuals, along the lines of the market-based tradable allowances model. Appurtenant groups might be defined according to categories of uses (for example, farming, residential water supply, commercial development, recreation). Such an arrangement would afford appurtenant groups a measure of management authority regarding their own usage. Freedom to trade and reallocate rights within groups could assist in the development of conservation measures and efficiency improvements. Appurtenant groups might also prove more effective at distributing water among members during times of scarcity and responding to fluctuations in user need and resource availability.

The existence of a larger market of permits might also facilitate intergroup coordination. At the same time, a centrally administered permitting scheme would preserve the ability of state administrators to set certain policy goals (such as equitable distribution), to monitor overall usage, and to ensure coordination of different user groups. My hope is that the examples of nested regimes and suggestions for eastern water offered in this Section will provide a starting point for the reassessment of regulated riparianism. Regulators should resist the tendency to call for greater centralization and regulatory control, and should question the assumption that ex ante decisions made by an administrative agency can alone solve the problems of scarcity and allocation that trouble eastern water today. The institutional insights gleaned from appurtenancy doctrine should inform the shape of water management institutions going forward.

To a large extent, this Note remains agnostic as to the appropriate degree of state oversight, noting only that administrators should be attentive to the institutional model adopted, as certain regulatory forms may themselves produce inefficiencies. Specifically, regulators might find that cross-group coordination would best be achieved not via more assertive forms of regulation, which have proved disastrous in numerous arenas, but rather by market incentives or the formation of larger umbrella institutions encompassing multiple smaller user groups. On the other hand, a more aggressive regulatory approach may be necessary where parochial community interests produce cross-group inefficiencies or inequities that the market alone cannot resolve. Precisely what balance should be struck between appurtenant groups and state oversight in the management of eastern water is a question the Note does not seek to answer, and one whose answer will very likely be contextual. Having argued for multilevel governance and having sketched a few possibilities, this Note recognizes that the remaining challenge is to understand exactly what institutional design—what allocation of decisionmaking, monitoring, and enforcement functions—will best achieve the goals of efficiency, equity, and conservation in water usage and distribution.

At a time when eastern states face continuing water shortages, and regulated riparian statutes are in need of reassessment, this Note offers new ideas about water management by turning to two different sources of inspiration: the common law and the growing body of literature on common property regimes. Despite the criticism it has garnered, the appurtenancy requirement exhibits certain durable principles for conserving scarce resources that this Note seeks to highlight. Moreover, in comparing appurtenancy to reasonable use, this Note provides a new theoretical perspective contrasting limiting and balancing rules—with which to assess resource management systems and the rules they impose.

The Note puts these conclusions to work in its analysis of regulated riparianism, concluding that the retention of a reasonableness inquiry may undermine efforts to conserve scarce water supplies. Rejecting the efficacy of such open-ended balancing tests, the Note instead urges consideration of a new institutional model—the nested enterprise model—as a way to manage scarce resources. In the context of eastern water, the nested enterprise model may offer policymakers a way to incorporate a reconceptualized principle of appurtenancy—a principle that attends to scarcity and respects the expertise of users in conserving common resources—into the broader oversight and authority of the state. While the doctrinal rule may have faded, the core concepts of appurtenancy can thus have continuing relevance. Embedding those concepts in our water law and policy will better allow us to conserve and manage our water supply.

## Stockwatering Exemption

Imagine a dairy farm in the arid eastern region of Washington State. The dairy's herd of 2500 cows consumes approximately 45,000 gallons of water per day (gpd). Water is a public resource in Washington, so prospective water users must apply for a permit from the State before withdrawing water from streams or aquifers. The permitting statute exempts specific types of small withdrawals from these permitting requirements, however, including "any withdrawal of public ground waters for stock-watering purposes...." Although other exempt uses are limited to 5000 gpd, the dairy farm believes that this limitation does not apply to its withdrawal of 45,000 gpd because the statute exempts "any withdrawal" for stock-watering purposes.

The Washington State Department of Ecology (Ecology), the Pollution Control Hearings Board (PCHB), and the Washington State Attorney General (Attorney General) recently examined this issue and arrived at conflicting conclusions. Ecology initially concluded that stock-watering withdrawals are limited to 5000 gpd by relying on the statutory and historical context of the exemption. In 2001 the PCHB agreed with this interpretation in DeVries v. Department of Ecology. A 2005 Attorney General opinion concluded that the plain meaning of the statute allows unlimited stock-watering withdrawals, however, and Ecology has subsequently changed its practices to accord with this opinion. The distinction between a limited and unlimited exemption is significant because declining groundwater levels and a proliferation of exempt groundwater withdrawals currently impede sustainable management of Washington's groundwater resources. Increasing reliance on groundwater has overstressed aquifers across the nation. In Washington, rapid population growth has placed significant pressure on groundwater resources.

These demands have caused declining groundwater levels across the state, especially in the eastern and southeastern regions, where the majority of the state's livestock farms are located. As water availability declines, Ecology is forced to deny an increasing number of water use applications to protect senior water rights. Consequently, new water users are turning to the statutory exemptions from the groundwater permitting requirements. Although the exact quantity of exempt withdrawals is unknown, because such withdrawals are unregulated, it is clear that the total amount of groundwater withdrawn under the exemption is significant. An unlimited exemption for stock-watering could exacerbate current groundwater management issues, because such exempt withdrawals occur outside Ecology's permitting authority. This Comment analyzes the conflicting interpretations of Washington's stock-watering exemption and argues that the exemption is limited to 5000 gpd. This conclusion is based on Washington's principles of statutory interpretation, the statute's historical background, and public policy considerations.

## Washington Water Law Comprehensively Regulates Groundwater

Washington groundwater is comprehensively regulated by a groundwater code that exempts small withdrawals from its permitting system. The Legislature enacted the groundwater code to prevent destructive competition for groundwater by providing uniform regulations and a centralized administration for governing water rights. Although specific types of withdrawals are exempt from the groundwater code's permitting requirements, they remain subject to other regulations within the groundwater code.

The prior appropriation doctrine governs water use in Washington, as it does in most western states. Under the riparian doctrine, which traditionally governed water use in the eastern states, only lands adjacent to water are entitled to a water right. In contrast, the prior appropriation doctrine allows water users to acquire water rights by appropriating water and putting it to a beneficial use. Prior appropriation also protects senior water users by providing that between appropriators from the same water source, the first in time has a higher priority water right. The gold miners and farmers who first settled the western territories developed this system because land that was not adjacent to water in the arid regions of the west was practically worthless under the riparian doctrine. Appropriation based on need, rather than on riparian property rights, facilitated successful settlement.

Between 1889 and 1917, Washington recognized water rights under both the riparian and prior appropriation doctrines. Beginning with Thorpe v. Tenem Ditch Co. in 1889, the Washington State Supreme Court considered the question of water rights in several cases and protected both rights by appropriation and rights by riparian ownership. In addition to the case law, Washington had many conflicting statutes regarding the right to use water. This resulted in a chaotic and confused state of the law, which, in turn, led to continuous disputes among water users throughout the state. To create clarity and security for water users, the Washington State Legislature enacted a comprehensive surface water code in 1917. This code adopted the prior appropriation doctrine and created a centralized administration for determining the existence of water rights and establishing new rights, thereby ensuring the most efficient use of water resources.

Several decades later, the 1945 Legislature passed the groundwater code as a supplement to the surface water code, extending the protections and regulations of the surface water code to groundwater. Although the Legislature did not explicitly state its purpose for enacting the groundwater code, the Washington Department of Conservation and Development (Department of Conservation), which was the agency responsible for water resources in 1945, described the reasons for its enactment in a 1946 report. In the report, the Department of Conservation stated that the groundwater bill was prepared and sponsored at the request of an association of Washington cities to protect the 126 cities that depended on groundwater for domestic uses, irrigation, and industrial purposes.

The Department of Conservation also noted that "destructive competition has been the experience of certain other states," and "in some sections of this State, uncontrolled withdrawal of [ground]water has already caused damage to existing rights and investments." This report indicates that the Legislature enacted the groundwater code to protect existing rights, avoid damaging competition, and provide security for future development.

### The Ground-water Code Provides Comprehensive Regulations

The groundwater code is supplementary to the surface water code. Because the groundwater code extended the provisions of the surface water code to govern groundwater appropriations, it is important to first understand the relevant provisions of the surface water code. The surface water code adopted the prior appropriation doctrine and created a permitting system that allows the state to control the distribution and utilization of surface water resources. Under this permitting system, a prospective water user must obtain a permit from Ecology before appropriating water. Before issuing a permit, Ecology must find "(1) that water is available, (2) for a beneficial use, and that (3) an appropriation will not impair existing rights or (4) be detrimental to the public welfare." The surface water code also mandates that none of its provisions "shall be construed to lessen, enlarge, or modify... any existing right acquired by appropriation, or otherwise." The groundwater code adopted most of these surface water code provisions. The groundwater code's purpose is "regulating and controlling the ground waters of the state." Because groundwater withdrawals can impact surface water resources in areas where surface water and groundwater are connected, the groundwater code protects existing surface water rights by mandating that surface water rights existing when the code was enacted "shall not be affected or impaired by any of the provisions of this supplementary chapter." The groundwater code also adopts the prior appropriation rules and permitting system established in the surface water code. Unlike the surface water code, however, the groundwater code provides an exception to the permitting system.

***The groundwater code's general permitting requirement provides:*** After June 6, 1945, no withdrawal of public ground waters of the state shall be begun, nor shall any well or other works for such withdrawal be constructed, unless an application to appropriate such waters has been made to the department [of Ecology] and a permit has been granted by it as herein provided....

This general rule requires prospective groundwater users to apply to Ecology for a permit before withdrawing public groundwater, and Ecology must make the same findings required in the surface water code before granting the permit application. This general rule is followed by an exception, in the form of an exemption provision, and two modifying provisions. The exemption provision states:

Except, however, That any withdrawal of public ground waters for stock-watering purposes, or for the watering of a lawn or of a noncommercial garden not exceeding one-half acre in area, or for single or group domestic uses in an amount not exceeding five thousand gallons a day, or as provided in RCW 90.44.052 [applicable only in Whitman County], or for an industrial purpose in an amount not exceeding five thousand gallons a day, is and shall be exempt from the provisions of this section, but, to the extent that it is regularly used beneficially, shall be entitled to a right equal to that established by a permit issued under the provisions of this chapter....

This provision exempts four categories of water use from the general permitting rule: stock-watering, irrigation, domestic uses, and industrial uses. Three of these categories are specifically limited-irrigation of a lawn or garden is restricted to a half acre, and domestic and industrial uses are limited to 5000 gpd-while the stock-watering provision does not contain a specific limitation. Although these categories of withdrawals are exempt from the permitting requirement, all groundwater withdrawals are subject to the mandate that existing surface water appropriations "shall not be affected or impaired by any provisions of [the groundwater code]," and the mandate that senior appropriators are entitled to a "safe sustaining yield" of groundwater.

***Two modifying provisions follow the exemption provision:*** Provided, however, that the department from time to time may require the person or agency making any such small withdrawal to furnish information as to the means for and the quantity of that withdrawal: provided, further, that at the option of the party making withdrawals of ground waters of the state not exceeding five thousand gallons per day, applications under this section or declarations under RCW 90.44.090 may be filed and permits and certificates obtained in the same manner and under the same requirements as is in this chapter provided in the case of withdrawals in excess of five thousand gallons a day. The first proviso authorizes Ecology to require exempt users to report the purpose and quantity of their withdrawals, and describes all exempt uses as "small withdrawal[s]." The second proviso, added in 1947, provides that users may voluntarily obtain a permit for their exempt withdrawals, and it describes the exempt withdrawals as "not exceeding five thousand gallons per day." When reviewing the statute as a whole, it is unclear whether the statute allows unlimited use because it lacks a specific limitation for the stock-watering exemption, or whether the statute limits stock-watering to "small withdrawals" not exceeding 5000 gpd through the modifying provisos.

While the groundwater code's stock-watering exemption lacks clarity, what is clear is that the circumstances surrounding the surface water and groundwater codes' enactment strongly indicate that the Legislature created these codes to minimize water disputes and protect the public's water resources from destructive competition. The groundwater code adopted the surface water code's protections and regulations for groundwater. Unlike the surface water code, however, the groundwater code contains an exception to the general permitting requirements for specific categories of water use, including withdrawals for stockwatering. This exemption statute fails to clearly convey the limitations that govern the stock-watering exemption.

# Bibliography


Alka Rani Upadhyay: *Aquatic Plants for the Waste Water Treatment*, Daya, Delhi, 2004.

Allan, J. A and Chibli Mallat: *Water in the Middle East: legal, political, and commercial implications*, I.B. Tauris, London, 1995.

Anderson, Mary P. and William W. Woessner: *Applied Groundwater Modeling: Simulation of Flow and Advective Transport*. Academic Press, San Diego, 1992.

Asaithambi S. : *Economics of Ground Water Management in India*, Abhijeet, Delhi, 2008.

Asit K. Biswas and Cecilia Tortajada: *Appraising Sustainable Development: Water Management and Environmental Challenges*, Oxford University Press, Delhi, 2005.

Bagis, Ali Ihsan: *Water as an Element of Cooperation and Development in the Middle East*, Ankara, Ayna Publications, 1994.

Bhakar S.R. : *Ground Water Hydrology : Theory and Practice*, Agrotech, Delhi, 2009.

Bhaskara, Harry: *Ending the Daily Water Chore In Against All Odds,* Panos Publications. London. 1989

Bhave, P.R. and R. Gupta: *Analysis of Water Distribution Networks*, Narosa, Delhi, 2011.

Biswas Asit K. : *Integrated Water Resources Management in South and South-East Asia*, , Oxford University Press, Delhi, 2001.

Bourne, Peter G.: *Water and Sanitation: Economic and Sociological Perspectives* Academic Press. Orlando. 1984.

Boyer , J.S.: *Measuring the Water Status of Plants and Soils,* Academic Press, N.Y., 1995.

Brooks, D.: *Water: Local-Level Management*, Ottawa, International Development Research Centre, 2002.

Bullock, John and Darwish, Adel: *Water Wars: Coming Conflicts in the Middle East*, London, Victor Gollancz, 1999.

Cazmuri, Renato : *"Chilean Water Policy Experience"*, Agriculture and Natural Resources Department, World Bank, Washington, D.C. Processed, 1992.

Conway, D.: *Climate Change and Water Resources in the Nile Basin*, University of London, London, 1993.

Cullis, A. : *Rainwater Harvesting: The Collection of Rainfall and Runoff in Rural Areas,* London, U.K.: IT Publications, 1986.

D'Itri, Frank M. and Lois G. Wolfson: *Rural Groundwater Contamination.* Lewis Publishers, Chelsea, 1987.

Doneen L.D.: *Water Quality for Agriculture,* Department of Irrigation, University of California, Davis, 1964.

Elmusa, Sharif S.: *Water Conflict: Economics, Politics, Law and Palestinian-Israeli Water Resources,* Institute for Palestine Studies, Beirut, Lebanon, 1998.

Engelman, R., and P. LeRoy : *Sustaining Water Washington,* D.C.: Population Services International, New York, 1993.

Erik Nissen-Peterson : *Rainwater Catchment Systems,* UK: Intermediate Technology Publications, 1999.

Feder, Paul A.: *A Guidebook to Groundwater: Resources and Education Opportunities in the Great Lakes Region,* Great Lakes Commission, Ann Arbor, 1993.

Freeze, R. Allan and John A. Cherry: *Groundwater.* Prentice Hall, Englewood Cliffs, NJ. 1979.

Garfinkle, Adam M.: *War, Water, and Negotiation in the Middle East: the Case of the Palestine-Syria Border 1916-1923,* Tel Aviv University, Tel Aviv, 1994.

Ghosh, N.C. and K.D. Sharma: *Groundwater Modelling and Management,* Capital Pub, Delhi, 2006.

Gould, J., and E. Nissen-Petersen : *Rainwater Catchment Systems for Domestic Supply: Design, Construction and Implementation,* London, U.K.: IT Publications, 1999.

Gupta K.R.: *Water Crisis in India,* Atlantic Pub, Delhi, 2008.

Gupta, O.P. : *Water in Relation to Soils and Plants : With Special Reference to Agriculture,* Agrobios, Delhi, 2002.

Heath, Ralph C.: *Basic Ground-water Hydrology,* Denver, CO: U.S. Geological Survey, 1983.

Helen Bannayan: *Water Resources of Jordan: Present Status and Future Potentials,* Amman, Friedrich-Ebert-Stiftung, 1993.

Holubeshen, Moufida; Amery, Hussein A.: *Annotated Bibliography of Water in the Middle East,* University of Toronto, Toronto, 1995.

Husain, Ahmad : *Environment and Water Resource Management,* Sumit Enterprises, Delhi, 2006.

Jana B L : *Water Harvesting and Watershed Management,* Agrotech, Delhi, 2008.

Jat, M.L. and S.R. Bhakar: *Ground Water Hydrology : Theory and Practice,* Agrotech, Delhi, 2009.

Jee Chandrawati and Shagufta : *Rainwater Harvesting*, A.P.H. Pub, Delhi, 2010.

Kally, Elisha and Gideon Fishelson: *Water and Peace: Water Resources and the Arab-Israeli Peace Process*, Westport, CT, Praeger, 1993.

Kanmony, J. Cyril : *Drinking Water Management : Problems and Prospects*, Mittal Pub, Delhi, 2010.

Khan M A : *Water Resources Management and Sustainable Agriculture*, APH, Delhi, 2008.

LeRoy, K. : *Sustaining Water Washington*, D.C.: Population Services International, New York, 1993.

Libiszewski, Stephan: *Water Disputes in the Jordan Basin Region and their Role in the Resolution of the Arab Israeli Conflict*, Center for Security Studies and Conflict Research, Zurich, 1995.

Linda M. Welkom: *Groundwater Chemicals Desk Reference*, Chelsea, MI, 1990.

Lloyd Myers : *Handbook of Water Harvesting*, Washington D.C.: U.S. Dept. of Agriculture, Agricultural Research Service, 1983

Lowi, Miriam: *Water and Power: the Politics of a Scarce Resource in the Jordan River Basin*, Cambridge University Press, Cambridge, Eng, 1993.

Mahajan Gautam : *Evaluation and Development of Ground Water*, APH, Delhi, 2008.

Matthess, Georg : *The Properties of Groundwater*, John Wiley & Sons, New York, NY. 1982.

McGarry, M.G.: *Matching Water Supply Technology to the Needs and Resources of Developing Countries,* United Nations. New York. 1987

Meenu Bhatnagar: *Groundwater Management : Sustainable Approaches*, Icfai Books, Delhi, 2012.

Metcalf and Eddy Inc: *Waste-water Engineering: Treatment, Disposal, Reuse*, McGraw-Hill Book Co, New York, 1979.

Mishra Archana : *Water Harvesting : Ecological and Economic Appraisal*, Authors Press, Delhi, 2006.

Montgomery, John H. and Linda M. Welkom: *Groundwater Chemicals Desk Reference*, Chelsea, MI, 1990.

Murakami, Masahiro, *Managing Water for Peace in the Middle East : Alternative Strategies,* Tokyo and NYC, United Nations University Press, 1995.

Naff T. and Matsow R.: *Water in the Middle East: Conflict or Cooperation*, Boulder, 1990.

Nielsen, David M.: *Practical Handbook of Groundwater Monitoring*. Lewis Publishers, Chelsea, 1991.

Nissen-Petersen, E. : *Rainwater Catchment Systems for Domestic Supply: Design, Construction and Implementation,* London, U.K.: IT Publications, 1999.

Pacey A., and A. Cullis : *Rainwater Harvesting: The Collection of Rainfall and Runoff in Rural Areas,* London, U.K.: IT Publications, 1986.

Pahwa Prem S. : *Water Harvesting, Purification and Distribution Management*, Dominant, Delhi, 2001.

Palanisami, K : *Groundwater Management and Policies*, Macmillan Publishers India, Delhi, 2008.

Parikshit Ballabh: *Water Resource Management : Planning and Development*, Cyber Tech, Delhi, 2009.

Premjit Sharma: *Agricultural Drainage and Water Quality*, Gene Tech Books, Delhi, 2007.

Ranghaswami, M.V. ; K. Palanisami: *Groundwater Resources Assessment, Recharge and Modelling*, Macmillan Publishers India, Delhi, 2008.

Rao K. Nageswara : *Water Resources Management : Realities and Challenges*, New Century Publication, Delhi, 2006.

Rowe, Garry W. and Sylvia J. Dulaney: *Building and using a Groundwater Database*. Lewis Publishers, Chelsea, 1991.

Salameh, Elias and Helen Bannayan: *Water Resources of Jordan: Present Status and Future Potentials*, Amman, Friedrich-Ebert-Stiftung, 1993.

Sanghvi Sheela and Pahwa Prem S. : *Water Harvesting, Purification and Distribution Management* , Dominant, Delhi, 2001.

Sharma, R. K. and T.K.Sharma: *A Textbook Of Water Power Engineering*, S. Chand Publisher, Delhi, 2008.

Sharma, S.S.P. and U.H. Kumar: *Dynamics of Watershed Development and Livelihood in India*, Serials Pub, Delhi, 2010.

Singh A K : *Environment and Water Resources Management*, Adhyayan, Delhi, 2006.

Thapliyal, B K : *Democratisation of Water*, Serials Pub, Delhi, 2008.

Thomas, Harold E.: *The Conservation of Groundwater: A Survey of the Present Groundwater Situation in the U.S.*, McGraw Hill, New York, NY. 1951.

Tideman, E. M.: *Watershed Management – Guidelines for Indian Conditions*, Omega Scientific Publishers, Delhi, 1996.

Todd, David Keith: *Groundwater Hydrology*. John Wiley & Sons, New York, NY. 1959.

Van der Leeden, Frits: *Ground Water: A Selected Bibliography*, Port Washington, NY, 1974.

Walton, William C.: *Groundwater Resource Evaluation*. McGraw Hill, New York, 1970.

# Index

□□□